Mathematical Ecology

Mathematical Ecology

edited by
Peter Allen (Brussels), Werner Ebeling (Berlin),
Manfred Peschel (Berlin), Peter Schuster (Vienna),
Yuri M. Svirezhev (Moscow)

Mathematical Ecology deals with mathematical models of evolution processes, in the biosphere as a unity of growth and structure-building, with spring-up of new species, their interaction and possible extinction, with the impacts of human activities on the environment and the corresponding consequences for the biosphere.

The whole field seems to be a certain amalgam of suitable system philosophy, in which real world phenomena are considered through an ecological pair of spectacles with the help of system methodology and mathematics.

All publications published herein are of interdisciplinary interest for ecology, biology, economy and technical engineering.

Ecological and Demographic Consequences of a Nuclear War

by Yuri M. Svirezhev

Akademie-Verlag Berlin 1987

Author

Yuri M. Svirezhev

Co-Authors

G. A. Alexandrov
P. L. Arkhipov
A. D. Armand
N. V. Belotelov
E. A. Denisenko
S. V. Fesenko
V. F. Krapivin
D. O. Logofet
L. L. Ovsyannikov
S. B. Pak
V. P. Pasekov
N. F. Pisarenko
V. N. Razzevaikin
D. A. Sarancha
M. A. Semenov
D. A. Smidt
G. L. Stenchikov
A. M. Tarko
M. A. Vedjushkin
L. P. Vilkova
A. A. Voinov

Computer Center of the USSR Academy of Sciences, Moscow

ISBN 3-05-500193-1

Erschienen im Akademie-Verlag Berlin, DDR-1086 Berlin, Leipziger Straße 3—4

Lizenznummer: 202 · 100/409/86
Printed in the German Democratic Republic
Gesamtherstellung: VEB Druckerei „Thomas Müntzer“, 5820 Bad Langensalza
Lektor: Dr. Reinhard Höppner
LSV 1005
Bestellnummer: 763 642 8 (6975)
01800

FOREWORD

The post-war 40 years have seen unprecedented rates of scientific and technical progress, and it would be no gross overestimation to call the present time an epoch of scientific and technical revolution.

The arrival of new efficient technologies, capable of increasing labour productivity manyfold, of novel materials, and new energy sources, of new consumer goods, inspired great hopes that the scientific and technical revolution would drastically improve the quality of life of people by saving human labour, and providing more leasure for cultural perfection, to say nothing of improving the living standards.

However, time was passing by and we must admit that these hopes turned out to be far from being justified. And when hopes are replaced by frustration a negative effect crops up giving rise to pessimism verging on negation of the expedience of technical progress. It is even heard sometimes that scientific discoveries are pernicious.

But scientific discoveries are not the point. Any scientific finding by itself can neither be harmful, nor useful. Only its applications may be harmful or useful.

Perhaps this statement may be best confirmed by the discovery of applications of nuclear energy. On the one hand, it provides mankind with practically inexhaustable energy resources; on the other hand, military applications of this discovery threaten to exterminate mankind and all the living things on our planet. Therefore, it is the sacred duty of scientists to give an estimation of the implications of the nuclear war and to make the results widely known to the public. In our "seething" world this task becomes extremely critical. I would say it is a task of superpriority.

Today there are a lot of resolute opponents of the nuclear war, but there are quite many advocates of the nuclear war as the only remedy to save human society of its ailments. However, arguments of both are rather on the emotional than on the scientific side. They are based on qualitative factors, which are very numerous and contradictory; everything depends upon the choice of factors, thereby making the conclusions about the outcomes of the nuclear conflict pretty subjective.

Consequently, only a quantitative analysis, which takes account of all essential factors determining the dynamics of processes, – mechanical, physical, chemical, biological ones, – can give a realistic forecast of the nuclear war consequences. Though today not all the necessary initial data have yet been accumulated to enable us to obtain a unique solution to the problem of forecasting, still we already have the possibility to put our forecasts in a sufficiently narrow range between the minimal and maximal possible effects.

In the present study by professor Yu.M. Svirezhev and co-workers, the authors tried to account all the available body of physical, chemical, biological (especially radiobiological) evidence. This allows us to state that their forecast gives not only a qualitative picture of the processes triggered by the nuclear war, but also gives quantitative results that would not essentially differ from reality.

In future, as the information will be accumulated and made more accurate, the suggested techniques of quantitative estimations and mathematical models of various processes could be used to improve the forecasts. However, already now any sane person will come to the unique conclusion that the nuclear war is inadmissible.

A.A. DORODNICYN

Academician,
USSR Academy of Sciences
Chairman,
Soviet National Committee for SCOPE

ACKNOWLEDGEMENTS

This work was supported by SCOPE and the Soviet National Committee for SCOPE. The authors would like to thank Professor M. Harwell (Cornell University) and Professor T. Hutchinson (University of Toronto) for useful discussions. The authors express their sincere gratitude to Miss. L. Goering for her invaluable help in translating the manuscript from Russian, to Mr. A.P. Repiev for some text corrections, as well as to Mrs. L.R. Novototskaya, Miss T.M. Guseva, and Miss K.S. Baizhanova for the technical assistance in preparing the manuscript for publication.

CONTENTS

INTRODUCTION

Though the last two World wars were called World wars, their ecological consequences were of regional nature and only the emergence of nuclear weapons with the energy potential sufficient to change global biospheric processes resulted in the development of a situation when a war may cause global ecological consequences. If in the past wars were regional disasters, a nuclear war would be a global one.

In the course of the last forty years the ghost of a nuclear war has been hovering over humanity. Scientists in many countries have done a great deal to comprehend and assess the possible implications of a nuclear holocaust. The destruction of cities, the mass death of people during nuclear explosions, the radioactive contamination of the earth, water, and air and many other similar phenomena have been predicted. The forecasts were horrible, envisaging the destruction of civilization, a new "dooms-day". All this gave rise to a natural response, which manifested itself in the broad expansion of the anti-war movement throughout the world.

However, all this research had, in our opinion, two substantial faults, which haves skilfully been employed by a small but clever enough body of people – the "evil genii" of our civilization, who are defending the very idea of using nuclear weapons.

Firstly, these research works did not contain enough scientifically based quantitative evaluations of the longterm consequences of a nuclear war. They were of more emotional than scientific nature.

Secondly, they did not provide definitive forecast of the fate of human civilization after a nuclear war. The impression was formed, that a ramified network of anti-nuclear shelters, the storing up of strategic storage of food and other resources, which undoubtedly enables to diminish considerably the first immediate effects of nuclear strikes, would make it possible to survive a nuclear war. Moreover, all this created an illusion that a nuclear war might be won by one of the participants.

There exists yet another illusion – that of a limited nuclear war. Although, at present, an enormous amount of nuclear arms has been built up (some experts believe the world nuclear arsenal to near 40 000 Mt) some people think that in the future nuclear war it would be possible to resort to just a few nuclear strikes on military objects and then to sit at the negotiating table. It is difficult to understand what is more in such reasoning – supernaivety or supercynicism. Warfare specialists consider that the most efficient strategy in a nuclear war would be the strategy of using nuclear arms at the very first hours, even minutes of the war. And in so far as military objects and especially the launching devices are normally well camouflaged and include many false targets (e.g. empty silos), the blows would consequently be effected upon the cities, concentrating of the country's main industrial and demographic potential. After the nuclear strikes have been made, the escalation of a nuclear conflict can not be stopped.*

Until recently no account was taken of three factors of nuclear war, which, as is clear now, would cause climatic and ecological consequences on a global scale. These consequences would affect not only the territories of belligerent countries but all the rest of the world as well. Therefore, the conception of survival of the non-participating countries also becomes illusory.

The first factor is the "nuclear winter". No serious consideration was given earlier to the consequences affecting the climate, for usually an analogy was made between nuclear explosions and big volcanic eruptions. In both cases an enormous amount of aerosols is released into the atmosphere, thus, creating a kind of a screen and reducing the temperature of the lower layers of the atmosphere. It was, therefore, believed that the climatic consequences of a nuclear war and of large-scale eruptions would be similar. During the eruption of Sambor in Indonesia in 1914 over a 100 km^3 of dust and ashes were released into the air. This is by several orders greater than can be released if all nuclear stockpiles are exploded. The climatic consequences were significant: in the Northern hemisphere the summer of 1916 was cold and rainy, but not lasting and far from being disastrous.

However, as it is clear now, this analogy was erroneous. Nuclear strikes on cities would cause huge fires, resulting in emissions into the atmosphere of an amount of soot (it is important that

* President J.F. Kennedy advised his staff to read "The Guns of August" by B. Tachman, in which the escalation of the events and the decisions that led to the World War I have been very graphically shown. Although the capacity of military equipment has since increased immeasurably, we believe that the principal mechanism of escalation is described in the book quite adequately.

it is soot, rather than any other substance) enough to diminish the air transparency a million times. As a result the average temperature of the surface of the Earth in the Northern hemisphere would fall in the very first months by 20°–25 °C, in some areas even by 30°–35 °C. A similar picture, though somewhat less prominent would be characteristic of the Southern hemisphere as well. For a whole year our planet would plunge into a severe "nuclear winter" and a deep "nuclear night". Obviously, this global climatic disaster would also cause a global ecological catastrophe.

The second factor, strange as it may seem, is a consequence of scientific and technological progress, and the importance of this factor would increase in time. This is radioactive contamination resulting from nuclear strikes on nuclear power stations, nuclear fuel plants and nuclear arms factories, as well as storages of radioactive wastes, which would obviously be attacked among other energy-producing installations.

It should be noted that in earlier studies radioactive contamination was not among the main factors causing ecological disaster. This was due to the fact that radioactive fallout from so-called "clean" nuclear blasts, particularly if distributed over the total Earth surface, produces comparatively low values of radioactive contamination. It was only recently that we began to clearly understand that nuclear strikes on nuclear power stations, nuclear fuel plants, nuclear arms factories, and radioactive wastes storages result in radioactive contamination by long-lived (which is significant) isotopes of such intensity that vast areas would become radioactive deserts for many years to come. This entitles us to speak of nuclear strikes on nuclear power and industrial installations as of ecological weapons, capable of causing ecological catastrophes on areas of hundreds of thousands square kilometres, i.e. nothing less than a global catastrophe.

The third factor is the contamination of the environment – increased many times – by such conventional pollutants as oil, nitrogen and sulphur oxides, heavy metals, various toxic substances such as pesticides and herbicides, resulting from the distruction and burning of industrial enterprises and of large scale forest fires. Moreover, in so far as scientific and technological progress enables nuclear, chemical, mineral extracting and other industries to grow fast enough, the two latter factors become still more dangerous for the environment in a nuclear conflict.

These and other factors of a nuclear war taken separately and, which is particularly important, in terms of their synergetic effect, require to consider afresh the consequences of a nuclear war, which hitherto appeared clear. As a result of our analysis we can assert that a nuclear war would be disastrous not only in its demographic and economic effects for the immediate participants of the conflict but would also cause a global climatic and ecological disaster, i.e. a global biospheric disaster. The nuclear arms are global biospheric arms.

One cannot categorically assert that a nuclear war would destroy the planet's total biosphere. However, it would switch on the starting mechanisms of "nuclear successions", would change the entire biogeography of the planet and the biosphere would enter a new state. The possibility itself of the existance of Homo sapiens as a species as well as the possibility of preservation of its ecological niche in this new biosphere would become rather questionable.

A few words about the structure of the book. We isolate the following factors leading to degradation of ecosystems after a nuclear war:

1. "Radiation shock". 2. Fires. 3. "Nuclear winter" and "Nuclear night". 4. Radioactive contamination of soil and land surface water. 5. "Conventional" pollution, increased many times. 6. Increasing UV-radiation after the termination of the nuclear winter.

Accordingly, the effects of each of these factors on the ecosystems are discussed in a separate chapter. An attempt has been made to provide quantitative evaluations of the respective effects and of a qualitative description of their synergetic actions. Where it is possible, the principle of biospheric description is used, long-term effects of the nuclear war factors on various bioms are assessed and their evolution, i.e. nuclear successions are discussed. The effects of these factors on freshwater ecosystems are also considered.

Besides purely ecological effects we also consider demographic effects, namely, destruction of the system of providing the population with foodstuff, somatic and genetic effects of nuclear war. In the description of the evolution of the biosphere after nuclear war we widely use the mathematical model of global biospheric processes, developed by us.

In our assessments, besides special literature we used various reference books and geographical atlases as well as various books on ecology and biogeography. No references in such cases were usually made.

1. SCENARIO OF NUCLEAR WAR. CALCULATIONS OF DOSE FIELDS AND RADIOACTIVE CONTAMINATION

1.1. Scenario of nuclear war

In our analysis of the possible ecological consequences of nuclear war and its effect on global biospheric processes we used the AMBIO scenario [79]. The authors of the scenario assumed that a nuclear conflict would occur between two main opposing sides and that the nuclear detonations would occur almost immediately. Less than half of the total nuclear arsenal of the USSR or the USA would be used in the war. The total nuclear charges used by both sides would amount to 5742 megatons. Attacks would affect all of Europe, the USSR, North America (USA and Canada) and the region of the Far East including Japan and South Korea. It was assumed that attacks would also occur in Third World countries not directly taking part in the war, as a means of disrupting their economic potential and diminishing their importance in the post-war situation. 14 744 targets would be struck, with 5569 Mt being detonated in the northern hemisphere and 173 Mt in the southern hemisphere. The attacks would fall on major cities and on industrial energy systems. Therefore, according to the conception of the AMBIO scenario, it can be accurately considered that the intensity of nuclear attacks would be proportionate to the population density in these regions. The hypothesis that the distribution of industrial and energy centers (including atomic industry and nuclear energy plants) is proportional to the distribution of population density is also completely likely.

In our further calculations we delineated three basic regions:

I. the USSR (European part, Caucasus, Urals, Central Asia)
II. Europe (including Turkey, excluding the USSR)
III. North America (the USA and the southern part of Canada).

Unfortunately the total nuclear yield of nuclear detonations falling on each region was not indicated in the AMBIO scenario. If we calculated the distribution percentages of targets in each region according to the attached maps (23 % – region I; 46 % – region II: 35 % on Western Europe and Turkey and 11 % on Eastern Europe; 31 % – region III) and if we distributed the total yield of nuclear detonations by region in proportion to these percentages, the basic principle of the AMBIO scenario of equivalence in an exchange between the two sides taking part in the nuclear conflict would not be adhered to.

Let us assume that the total yield of nuclear detonations in these regions would amount to approximately 5400 Mt. According to the AMBIO scenario the total yield would be 5569 Mt for the northern hemisphere and 173 Mt for the southern hemisphere. In view of the fact that the regions we have chosen do not include all targets in the northern hemisphere (but almost all have been considered), the estimate of 5400 Mt can be considered sufficiently accurate. If we distribute this amount by region in proportion to the percentage of targets, we then get the following estimate of the nuclear yield:

Region I – 1242 Mt;
Region II – 2484 Mt. 1090 Mt on Western Europe and Turkey and 594 Mt on Eastern Europe
Region III – 1674 Mt.

It is easy to see that with such a distribution the basic principle of the AMBIO scenario would not be adhered to.

Therefore we have used the simplest possible assumption that corresponds to this principle: 1800 Mt of total nuclear yield would fall on each region.

1.2. Method for calculating dose fields and radioactive contamination (without considering the destruction of nuclear power plants)

Within 24 hours after the explosion of a nuclear bomb of 1 Mt, the radioactivity of the material produced by the detonation would amount to 5.9×10^9 Ci – see figure 1. According to [22], about half the radioactive materials in a surface explosion would fall to earth in the form of radioactive fallout. The remaining part of radioactive materials would be emitted into the stratosphere, "spread" over the whole planet, and in a few years would fall to the earth's surface, but would already be much less radioactive. The relationship between possible surface and air explosions is not certain. In our calculations we assumed that most of the explosions would be surface explosions, and that 50 % of the total radioactivity would fall in the form of radioactive fallout.

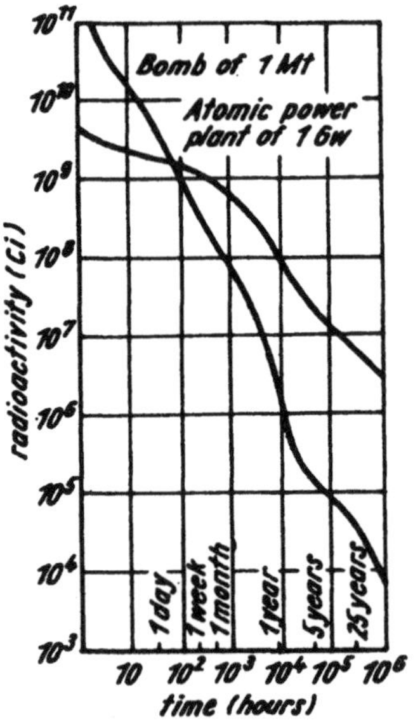

Fig. 1. Rates of radioactive decay of products of a "clean" nuclear detonation and products of atomic power plant destruction (from F. Burnaby, L. Kristeferson et al. Ambio, 11, N 2–3, 1982, p. 99)

Given a total nuclear yield of 1800 Mt on each of the three regions, the total radioactivity in a region is

$$P = 5.9 \times 10^9 \times 1800 \times 0.5 = 5.3 \times 10^{12} \text{ Ci}.$$

Let S_i be equal to the area of territory with a population density of n_i, the radioactive contamination of which is σ_i. Then according to our hypothesis of the proportionality between the distribution of nuclear yield and the distribution of population density,

$$\Sigma_i \sigma_i S_i = P, \qquad \sigma_i = k \cdot n_i \tag{1}$$

where k is the coefficient of proportionality, i is the number of gradation of population density. We consider three gradations with average density: $n_1 = 150$ ind./km², $n_2 = 30$ ind./km², $n_3 = 10$ ind./km². Obviously $\Sigma_i n_i S_i = N$, where N is the total population in the corresponding region. Then

$$\sigma_i = \frac{P}{N} \cdot n_i. \tag{2}$$

The total population in region I is 250 million, in region II – 530 million, in region III – 260 million. Given P(t), it is easy to calculate the density of radioactive contamination in each region for territories with different population densities.

a) *Calculation of dose for γ-radiation*

The intensity of γ-radiation for γ-quants with an energy of 0.7 MeV (average energy oт γ-quants in the products of a nuclear detonation) and with a density of contamination of 1 Ci/m² is 10 – 15 Roentgen (R)/hour (on the level of 0.9 m above the smooth surface) [22]. In taking into account the influence of microrelief the correction coefficient is included. This coefficient for virgin flat land is equal to 0.8 [28]. For a less smooth relief with isolated forests this coefficient is equal to 0.5.

Given this value of the coefficient and an average value of dose intensity, we get the relationship 1 Ci/m^2 – 6.25 R/hour.

Knowing the density of contamination at a given moment, we can use it to calculate the intensity of γ-radiation above the surface of the earth.

The radioactivity of products of a nuclear detonation changes over time according to the dependence [22] (time is measured in hours)

$$P(t) = P(1)t^{-1.2} \quad (3)$$

Let us determine the coefficient P(1) in the following way. According to figure 1 P(24) = 5.87×10^9 Ci for 1 Mt. Then P(1) = $5.87 \times 10^9 \times 24^{1.2} \approx 2.6 \times 10^{11}$ Ci.

Obviously the density of contamination $\sigma_i(t)$ varies according to the same law as P(t). Given $\sigma_i(t)$, it is possible to find the corresponding intensity of dose at a level of 0.9 m above the earth's surface. If we substitute (3) for (2) we get the following formula of dose intensity of γ-radiation in territories of different population density (for each region):

$$J_i(t) = J_i(1)t^{-1.2} = 2.6 \times 10^{11} \times 6.25 \times 1800t^{-1.2}\,(n_i/N) \times 10^{-6} = 2.9 \times 10^9 t^{-1.2} \times (n_i/N). \quad (4)$$

Here n_i is the population density, N is the total population in the region, the coefficient 10^{-6} is included for scaling since the contamination density is measured per square meter but the population density per square kilometer.

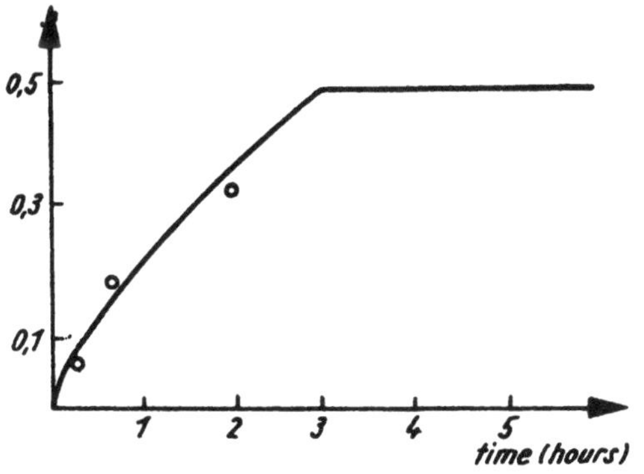

Fig. 2. Dynamics of sedimentation of radioactive fallout (in per cent)
o – experimental data

We will be interested in doses accumulated during a day and a year. In calculating the day doses it is necessary to take into account that the radioactive fallout will not settle immediately after the explosion: during the first 22 minutes, 4 % of the total radioactive material will settle in the largest particles. In the next 20 minutes 13 % more will fall; from 42 minutes to 2 hours 14.5 % more [26]. Let us approximate the initial part of the fall curve, r(t), by the dependence $r(t) = 0.2 \times t^{0.85}$ ($t < 3$ hours). Since we suppose that 50 % of the total radioactivity will be emitted into the stratosphere, we choose r(t) in the form (see figure 2)

$$r(t) = \begin{cases} 0.2t^{0.85}, & \text{if } t < 3 \text{ hours} \\ 0.5, & \text{if } t > 3 \text{ hours.} \end{cases} \quad (5)$$

Using formulae (4) and (5) we can calculate the dose of γ-radiation accumulated in one day for the regions and territories with different population densities being considered:

$$D_i(24) = \int_0^{24} r(t)\,J_i(t)dt = 3.7 \times 10^9 \times (n_i/N). \quad (6)$$

Table 1. The distribution of accumulated doses and the dose intensity (one year later) in regions and territories with different population densities

Accumulated doses (rad)	Regions	USSR I		Europe II		N. America III	
	Total population (million)	250		530		260	
	Powers of APP (Gw)	37		130		183	
	B \ A	β	γ	β	γ	β	γ
One day later (only bombs)	150	17750	2220	8875	1050	17075	2150
	30	3550	440	1675	200	3400	430
	10	1200	150	550	70	1150	140
One year later (only bombs)	150	30200	3775	14250	1775	29025	3650
	30	6050	750	2850	350	5800	725
	10	2000	250	950	120	1925	250
One year later (only destructions of APPs)	150	1450	180	2400	300	6850	850
	30	300	35	475	60	1375	170
	10	100	12	160	20	460	60
The total dose one year later	150	31650	3950	16625	2075	35900	4500
	30	6325	800	3325	420	7175	900
	10	2100	260	1100	140	2400	300
The dose intensity on the Earth surface one year later (rad/day, only destructions of APPs	150	8.1	0.7	13.5	1.1	38.5	3.2
	30	1.6	0.14	2.7	0.22	7.7	0.64
	10	0.5	0.05	0.9	0.07	2.5	0.21

A – types of radiation
B – population density (ind./km^2)

As a result we get, for example, for region I the following value of day doses of acute γ-radiation: for territories with a population density of 150 ind./km^2 – 2 220 rad; 30 ind./km^2 – 440 rad; 10 ind./km^2 – 150 rad.

Doses for the other regions are calculated by the same method. These are all contained in Table 1 (suitably rounded off).

Note that these doses are more correctly measured in Roentgens, but for consistency they will be measured in rads.

Let us now calculate the doses accumulated during one year:

$$D_i(8760) = D_i(24) + 2.9 \times 10^9 \times 0.5 \times (n_i/N) \times \int_{24}^{8760} t^{-1.2}\,dt = 6.38 \times 10^9 \times (n_i/N)$$
$$\cong 1.7 \times D_i(24) \qquad (7)$$

b) *The calculation of dose of β-radiation*

The calculation of dose of β-radiation in each concrete case demands an individual approach, taking into account size, geometry and position in space of the radiated objects. However, we will simplify this problem by supposing that radioactive fallout is distributed uniformly on biological objects with a high absorptive capacity (for example, along the canopies of stands of trees of high quality). Estimates of the intensity of the accumulated dose for a many-layered forest community, in which the canopy absorbs and keeps up to 90 % of radioactive fallout, are contained in book [71].

Given a contamination of $\sigma_e = 1 \times 10^{-3}$ Ci/m^2 the intensity of accumulated dose in points of growth of assimilating tree organs is about 2 rad/day at the initial moment. The intensity of accumulated dose of γ-radiation is 0.25 rad/day; it remains almost constant along the entire length of the tree.

In taking account of these data, we consider that the accumulated doses of β-radiation differ from that of γ-radiation by a coefficient of 8. It must be remembered that the doses of β-radiation are accumulated mainly by the plant cover. The corresponding data on doses of β-radiation are contained in Table 1.

For meadow and steppe communities (grassland communities) the dose of β-radiation accumulated by plants would be higher.

If we suppose that the depth of the plant cover is 20 cm and the radioactive contamination is uniformly distributed along the entire volume of plant cover, and the migrating nuclides are concentrated in the upper layer of soil at a thickness of 0.5 cm, then the following estimations of accumulated doses were done for a mixture of ten day old radioactive materials:

Given a contamination density of 1×10^{-3} Ci/m^2 the accumulated dose on the upper boundary of plant cover was 3 rad/day, in the middle part – 3.5 rad/day, on the soil surface – 2 rad/day [71]. About 0.25 rad/day of these doses are received in the form of γ-radiation. Note that this value is constant on all levels. Then, in using "γ-doses" to calculate "β-doses" for grassland communities the scale coefficient is 11–13.

1.3. Calculations of dose fields and radioactive contamination taking into account the destruction of atomic power plants

The destruction of atomic power plants and factories of nuclear fuel and ammunition leads to additional radioactive contamination. For calculation we again use the graphic in figure 1. In approximating this function we find (for atomic power plants of 1 Gw):

$$P_a(t) = \begin{cases} 4 \times 10^9\, t^{-0.225} & \text{if } t \leqslant 4.6 \times 10^2 \text{ hour} \\ 10^{11}\, t^{-0.75} & \text{if } t > 4.6 \times 10^2 \text{ hour} \end{cases} \tag{8}$$

We consider the radioactive contamination caused by the destruction of atomic power plants only in terms of the dose accumulated over a year. We use the same hypotheses as above: 1. Immediately after the explosion only 50 % of the radioactive materials fall in the form of radioactive fallout; 2. radioactive fallout is distributed in proportion to population density.

A third hypothesis has been added: 3. Since the products of the destruction of atomic power plants contain mainly large particles, the sedimentation (with the exception of the half that is emitted into the stratosphere) occurs practically immediately.

For these calculations we need information on the number of atomic power plants and their total power in the regions considered.

According to the prognosis for 1985 [33], in the United States and Canada, there will be 193 plants with a total power of 183 Gw. In Europe (excluding the USSR) their total power will be 130 Gw. In the USSR according to [3], in 1985 there will be 21 atomic power plants with a total power of 37 Gw.

The calculations that follow were made according to the algorithm in the above subchapter, but in place of the value of radioactive contamination from one bomb of 1 Mt and the corresponding megatonage, we use a value for the destruction of an atomic power plant with a power of 1 Gw and the corresponding total power in each region, and instead of the formula for the diminishing of radioactivity $\sim t^{-1.2}$, we use formula (8). The results of the calculations of corresponding doses are contained in Table 1.

In addition to doses, Table 2 contains estimates for the density of radioactive contamination from nuclear bombs and from destructions of atomic power plants. In calculating the density of

Table 2. The distribution of radioactive contamination in regions and territories with different population densities

Density of radioactive contamination (Ci/m^2)	Regions / Total population (million) / A / B	USSR 250 I	Europe 530 II	N. America 260 III
	A	37	130	183
One day later (only bombs)	150	3.2	1.5	3.0
	30	0.65	0.3	0.6
	10	0.2	0.1	0.2
One day later (only destruction of APPs)	150	$22\cdot10^{-3}$	$37\cdot10^{-3}$	$105\cdot10^{-3}$
	30	$4.4\cdot10^{-3}$	$7.4\cdot10^{-3}$	$21\cdot10^{-3}$
	10	$1.5\cdot10^{-3}$	$2.5\cdot10^{-3}$	$7\cdot10^{-3}$
One year later (only bombs)	150	$3.8\cdot10^{-3}$	$1.8\cdot10^{-3}$	$3.6\cdot10^{-3}$
	30	$0.8\cdot10^{-3}$	$0.4\cdot10^{-3}$	$0.7\cdot10^{-3}$
	10	$0.3\cdot10^{-3}$	$0.1\cdot10^{-3}$	$0.2\cdot10^{-3}$
One year later (only destruction of APPs)	150	$2.7\cdot10^{-3}$	$4.5\cdot10^{-3}$	$12.8\cdot10^{-3}$
	30	$0.55\cdot10^{-3}$	$0.9\cdot10^{-3}$	$2.6\cdot10^{-3}$
	10	$0.2\cdot10^{-3}$	$0.3\cdot10^{-3}$	$0.9\cdot10^{-3}$
The total density of contamination one year later	150	$6.5\cdot10^{-3}$	$6.3\cdot10^{-3}$	$16.4\cdot10^{-3}$
	30	$1.3\cdot10^{-3}$	$1.3\cdot10^{-3}$	$3.3\cdot10^{-3}$
	10	$0.4\cdot10^{-3}$	$0.4\cdot10^{-3}$	$1.1\cdot10^{-3}$

A – power of an APP (Gw)
B – population density (ind./km^2)

contamination one year after the nuclear explosions we use Figure 1. The main portion of this contamination is caused by products of the destruction of atomic power plants, but the effects from nuclear bombs are also noted. However, the diminishing of radioactive contamination from bombs occurs much more quickly than contamination from products of the destruction of atomic power plants. Thus the intensity of doses contained in Table 1 (one year after the conflict) refer only to the radioactive contamination from atomic power plants. The intensity of doses of β-radiation are calculated for the level on the surface of the soil. In this case the intensity of accumulated dose is 2.9–3.1 rad/day (for 1×10^{-3} Ci/m^2), of which 0.25 rad/day is a "γ-dose" [71].

1.4. Distribution of doses and radioactive contamination in the regions of a nuclear conflict and the territories with different population density

In order to get the geographical distribution of doses and contamination, we used cartographic information. There are maps of population density with gradations: less than 1 ind./km^2, 1–10 ind./km^2, 10–50 ind./km^2 and above. But in the regions considered, the territories with a density of population above 200 ind./km^2 comprised an extremely small percentage of the total territory. For example, in Europe the area of territories with a density of 200–600 ind./km^2 is 2.6 %, and the area of territories with a density of 50–200 ind./km^2 is 54 %. In the other two regions the percentage of area with a density of 200–600 ind./km^2 is much smaller. Therefore we will include these territories in the gradation of 50–200 ind./km^2, shifting the mean density correspondingly. The territories with gradations of less than 1 ind./km^2 and 1–10 ind./km^2 will be combined into one gradation of 0–10 ind./km^2.

How will we evaluate the bias of the mean in the territories with a gradation of 50–200 ind./km^2? In these territories urban population is prevalent. Using the mean percentage of the urban population in developed countries (70 %), we calculate the weighted value of the mean for this gradation: 50 X 30 % + 200 X 70 % = 155. Therefore we use a value of 150 ind./km^2 as a mean population density in these regions.

How will the mean density be calculated for the other gradation?

In territories with a gradation of 10–50 ind./km^2 the percentage of urban and rural population is approximately equal. Therefore for these territories it is natural to use a value of 30 ind./km^2 as a mean density.

Finally, territories with a density of 0–20 ind./km^2. We will do the following. The total area of the USA (excluding Alaska) and the southern part of Canada is approximately 10^7 km^2. Approximately 260 million people live in this area. The territories with a population density of 50–200 ind./km^2 make up ~7.5 % of this total area; the territories with a density of 10–50 ind./km^2 make up 30 %. Thus, for the conservation of the total population balance the mean density of population in the remaining territory must be equal (rounded off reasonably) to 10 ind./km^2.

The total area of Europe (excluding the USSR, Iceland and Spitzbergen) and Turkey is 5.55 million km^2. Approximately 530 million people live in this area. Territories with a population density of 50–200 ind./km^2 (and above) make up 56.6 % of the total area; territories with a density of 10–50 ind./km^2 make up 28 %. Again we find that for the conservation of population balance we must get a population density of 10 ind./km^2 for the remaining territory.

We get the same estimations for region 1.

Then we use standard maps of distribution of population density in the regions considered, dividing the territories by the corresponding gradations. Now using the information contained in Tables 1 and 2, it is possible to calculate the accumulated doses and radioactive contamination density in each of the territories of the regions I, II, and III (see Figure 3).

1.5. Conclusions

Naturally, there would not exist such sharp boundaries of doses and radioactive contamination between territories of different population density – because of the continuous "diffusion" of radioactivity, the boundaries of the territories must be "fuzzy".

Naturally, because of transfer by winds and because of leaching from the surface, a part of the radioactive contamination will move to other regions, thus reducing the amount in the regions considered.

However, we consider that the proposed parametrization is sufficient for estimating the strong effects (such as the effects of a nuclear war). If the estimations are raised too high, they will be raised in relation to a certain scenario. And if we changed the scenario, e.g. only raising the total yield of nuclear weaponry, our estimations would be too low.

There is one further consideration in favour of our parametrization. It is practically impossible to foresee the initial conditions at the moment of the beginning of a nuclear war in their entirety. Neither we know an accurate distribution of attacks and their yield, nor an accurate distribution of targets of attack, nor an accurate forecast of the meteorological situation at that moment. Only the general assumptions that we used as a basis for our model of the distribution of doses and radioactive contamination are more or less beyond doubt. The same model can be used to estimate the distribution of "ordinary" pollution.

Finally, a few words about our assumption that equal nuclear yields would fall on each region. The parametrization we used here allows us to calculate easily the distribution for other scenarios. For example, if the total yield of nuclear detonations for a certain region is a factor of k, it is only necessary to change the accumulated dose and radioactive contamination (from bombs) in this region by a factor of k.

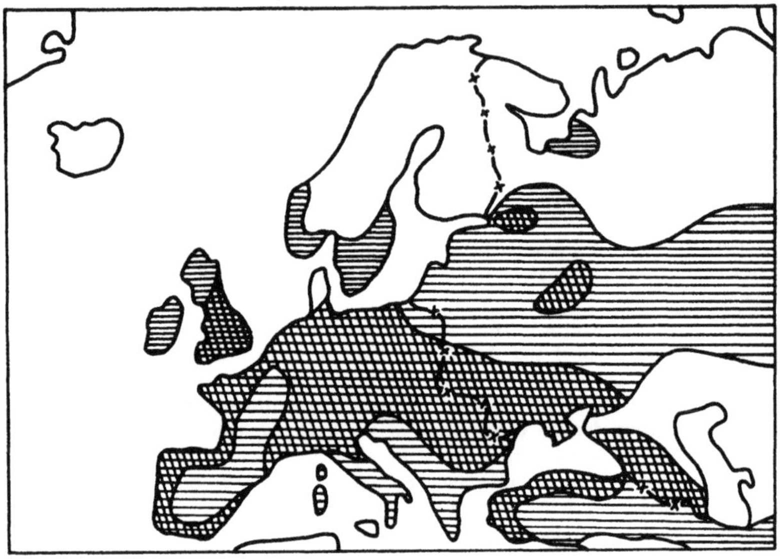

a)

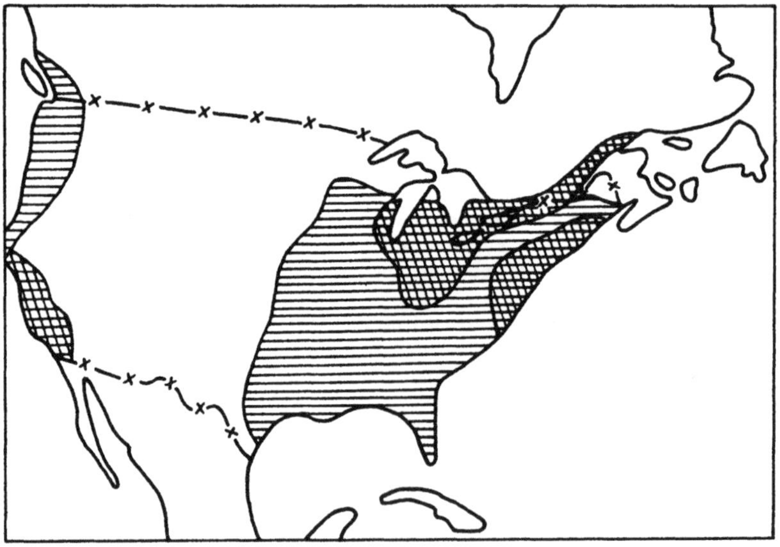

b)

Caption to the figure

Fig. 3. The distribution of doses and the distribution of radioactive contamination in the regions of Europe (a) and North America (b). The destruction of atomic power plants was considered

Regions	A	USSR I		Europe (excl. USSR) II		North America III	
		β	γ	β	γ	β	γ
Dose accumulated during the following day (rad)		17750	2220	8375	1050	17075	2150
		3550	440	1675	200	3400	430
		1200	150	550	70	1150	140
Dose accumulated during the following year (rad)		31650	3950	16625	2075	37900	4500
		6325	800	3325	420	7175	900
		2100	250	1100	140	2400	300
Radioactive contamination by long-lived isotopes one year later		$2.7 \cdot 10^{-3}$		$4.5 \cdot 10^{-3}$		$12.8 \cdot 10^{-3}$	
		$0.55 \cdot 10^{-3}$		$0.9 \cdot 10^{-3}$		$2.6 \cdot 10^{-3}$	
		$0.2 \cdot 10^{-3}$		$0.3 \cdot 10^{-3}$		$0.9 \cdot 10^{-3}$	

A – type of radiation

Of course, the estimates presented above, which are maximal, have a certain finite probability. Taking into account that multiple nuclear detonations would take place in the air rather than on the surface, that a part of the radioactivity would be carried to the ocean and sparcely populated territories, etc., our estimates of course should be diminished. However, we should keep in mind that averaged estimates, which are good, say, for insurance business, are unacceptable in the analysis of disastrous situations.

2. "RADIATION SHOCK" AND RADIOACTIVE CONTAMINATION THE IMPACT ON TERRESTRIAL ECOSYSTEMS

2.1. Introduction

Although ionizing radiation is always one of the factors of the environment of the earth, the doses it causes are so small (~0.1 rad/yr), that they do not have a significant influence on living organisms. However, after nuclear explosions the dose of ionizing radiation accumulated in the course of a day would be thousands of rads in the affected regions. Although approximately half the total radioactivity would be thrown into the stratosphere, the other half (in the event of surface explosions) would settle to the surface of the earth in the course of 3–5 hours, absorbed mainly by the plant cover. As a result of inertial sedimentation, plant leaves would be covered by radioactive products; this means that an even larger dose of β-radiation would be added to the enormous dose of γ-radiation (see Chapter 1). The plants (and of course the animals that live in the plant communities) would experience a "radiation shock" of unbelievable strength.

The fate of survivors would not be much better. Radioactive fallout consisting of long-lived isotopes (formed mainly by the destruction of atomic power plants and atomic industry), falling directly on the soil or leached into it from plants, would provide chronic irradiation of both plants and animals. The constant impact of this factor would continue for decades.

In this chapter, using the data on the impact of acute and chronic irradiation of plant communities and animals, and calculations of dose fields and radioactive contamination contained in Chapter 1, we will try to estimate the degree of impact of these factors of nuclear war on ecosystems and to evaluate the possible scale of destruction and degradation.

2.2. Comparative radiosensitivity of various organisms and communities

Various species of organisms differ greatly in their reactions to various doses of radiation. The doses accumulated by an organism over a short period of time (minutes or hours) are called acute doses (e.g. the irradiation causing them), as opposed to the chronic sublethal doses (and chronic irradiation), which an organism can accumulate in the course of its life. In this report we introduce the following classifications: irradiation acting in the course of the first day – acute; irradiation acting in the course of a year – acute-chronic; irradiation acting after a year – chronic.

Mammals are most radiosensitive and microorganisms are most radioresistant. Seed plants and lower vertebrates lie in between insects and mammals in radiosensitivity. Among plants, conifers are the least radioresistant. Dividing cells are most sensitive to radiation; therefore any component of the ecosystem (individual or population) is most susceptible during the growth process.

Although information on the radiosensitivity of individual plants and animals is necessary for a general description of the processes of radiation damage of ecosystems and communities, that alone is insufficient, since a complex system of multivarious secondary reactions plays an important role in these processes.

The increased development of more radioresistant species which share the ecological niche with species that are more sensitive to radiation (for example, the development under the canopy of phototropic plants after radiation damage to the tree layer) is a typical example of these reactions. The same is true for species in relationships of the type "prey-predator" or "host-parasite" (for example, weakening of trees is accompanied by an increase in the number of insect pests).

The degree of radiation damage is also largely dependent on abiotic factors (conditions of temperature, water, light and so on). However it can be said with certainty that the influence of other negative factors (pollution, damage to plants, non-optimal weather conditions, etc.) always increases the degree of radioactive damage. Consequently, in considering the impact of this factor we must always take synergistic interaction into account.

For the illustration in Figure 4 we have used data on the comparative radiosensitivity of various animal and plant communities (the meaning of the phrase "moderate degree of damage" will be explained below). It is apparent that coniferous forests are most sensitive to the radiation factor, while moss-lichen communities are least sensitive.

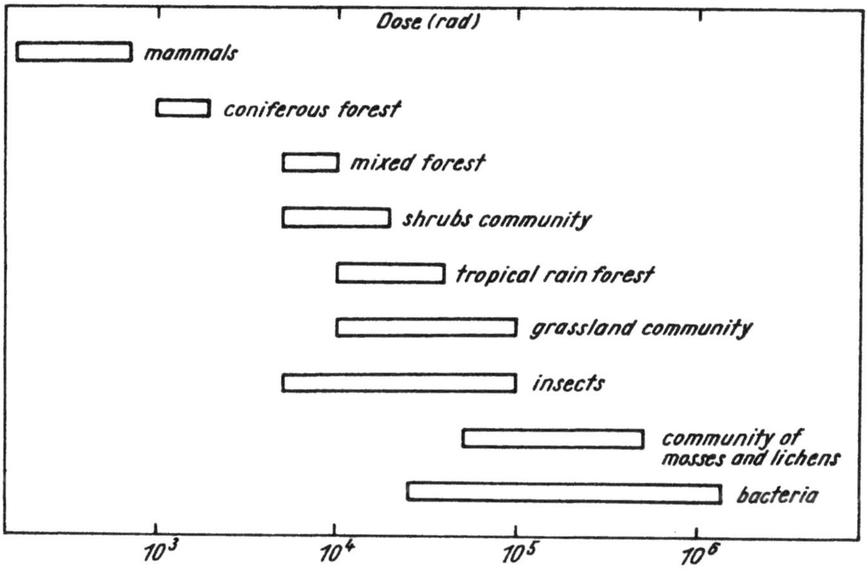

Fig. 4. Comparative tolerance of animal populations and plant communities to acute dose of γ-radiation. Indicated doses kill 50 % of the total population or damage moderately the plant communities

Probably we will need more detailed information on the radiosensitivity of communities. Thus, as a supplement to the drawing we can use Table 1, taken from [1].

Since forest ecosystems are most sensitive, and, on the other hand, since the main nuclear attacks would occur in the territory of North America and Europe, where coniferous, mixed and deciduous forests are most widespread, we will confine our analysis of radiation damage exactly to these plant communities.

Table 1. Classification of plant communities according to their radiosensitivity

Type of community	Doses (10^3 rad)		
	low degree of damage	moderate degree of damage	high degree of damage
coniferous forest	0.1–1	1–2	>2
mixed forest	1–5	5–10	10–60
tropical rain forest	4–10	10–40	>40
shrub community	1–5	5–20	>20
grassland community	8–10	10–100	>100
moss and lichen community	10–50	50–500	>500
lichen community	60–100	100–200	>200

2.3. Radiation damage of forest communities — I. Acute irradiation

Acute irradiation is conditioned mainly by the detonation of a nuclear bomb, both as a result of radiation directly after the explosion (the penetrating radiation acts within 10–15 seconds), and as a result of decay of short-lived isotopes after sedimentation of radioactive dust on the crown of the trees (first few hours after the explosion). It can be concluded that mainly the upper layer of the forest, needles and leaves, suffer from acute irradiation. For a given total dose the radiation damage will be strongest under conditions of single acute irradiation. This can be seen from the data in Table 2 [66a].

Table 2. Lethal doses (LD_{100}) of single acute (10–15 sec.) irradiation (γ-radiation) of forest stands during a vegetation period

Tree species	Doses (Roentgen)
Yew-tree	80
Pinus strobus	100
Picea glauca	102
Larix japonica	1225
Oak	800
Birch	800
Maple	1000

The death of forest planting is a result of the cessation of growth and the drying up of trees. The effect of sublethal doses is manifested in the retardation of growth, slowing of development, and the lowering of germinating capacity. The effects of the impact of acute irradiation as a function of dose are described in Table 3, taken from [66a].

Note that the relative character of radiation damage is more or less identical for various species and is the same for other types of irradiation — only LD_{100} differs.

The area damaged by penetrating radiation is relatively small, but the movement of the radioactive cloud, its "diffusion" and sedimentation of radioactive dust on the crown of trees (here we should also take into account the accumulated dose of β-radiation) increases the area of damage many times. For a "standard" situation (a surface explosion of a bomb of 1 Mt, wind velocity 24 km/hour) we get the following values of doses accumulated by the crowns of trees in the course of one day after the explosion (see Figure 5).

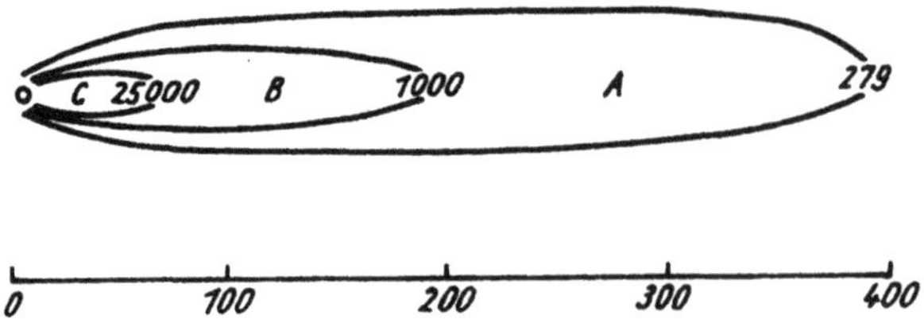

Fig. 5. Distribution of doses accumulated by crowns of trees one day after an explosion of a 1 Mt-bomb (wind velocity 24 km/h). Numbers 25000, 1000 and 275 point out values of the accumulated doses on corresponding lines

● – point of explosion

Table 3. Degree of damage of forests stands as a function of dose (for single acute γ-irradiation)

Dose (% of LD_{100})	Degree of damage (reactions)
10	Normal growth
25	Reduction of growth by 10 %
30	Reduction of growth by 30 %
35	Reduction of growth by 50 %
40	Reduction of growth by 60 %, pollen sterility
45	Reduction of growth by 60 %, pollen sterility, retardation of formation of generative organs
50	Same as above plus reduction of germinating capacity by 50 %, reduction of growth by 80 %
60	Same as above plus a sharp reduction of growth (up to 90 %), drying of parts of crown, dying off of trees with thin stems, death of young trees, an inviability of seeds
65	Same as above plus death up to 30 % of grown-up tree, cessation of growth of the others
70	Same as above plus death up to 45 % of trees
80	Same as above plus death up to 60 % of trees
90	Same as above plus death up to 75 % of trees
95	Same as above plus death up to 90 % of trees
100	Death of 100 % of trees (for deciduous trees a vegetative renewal is possible)

The close similarity of the reactions of trees to irradiation of γ-quants and β-particles makes it possible to use the information we have on the direct radiobiological impact of γ-radiation and the data contained in Table 3.

If the irradiation occurs over the course of a day, then LD_{100} for conifers (in a vegatation period) is 2000 rad [71]. In a rest phase LD_{100} is approximately 1.6 times greater than in a vegetation phase [71]. For deciduous trees LD_{100} is 5–10 times greater than for conifers.

If we take into account all of the above, we can conclude that if the nuclear attack occured during the summer, after one day all the trees in zone C (area ~1000 km^2) would perish; in zone B (area ~4000 km^2) only conifers would be damaged (moderate degree of damage).

But the damaging action of the radiation factor is not limited to this.

2.4. Radiation damage of forest communities — II. Acute-chronic irradiation

Acute-chronic irradiation is determined by the radioactive contamination from radioactive particles falling on the crowns of trees, some of which are then leached onto the surface of the earth. Since the time of dose accumulation of acute-chronic irradiation is one year, a substantial contribution to the total contamination is also made by the products of the destruction of atomic power plants (see chapter 1).

Vertical and horizontal migration of radioactive substances begins immediately after the fallout on the crowns. This is conditioned by several natural factors. Rain washing over the crowns of the trees transfers radionuclides from the upper parts of the crowns to the lower, and then below the canopy of the forest. Wind carries radioactive substances from one crown to another, as well as partially below the canopy. A part of the radionuclides is transferred to the litter together with leaf fall. As a result, a year later the amount of radionuclides in the crowns of trees would be several times lower (in deciduous forests); radioactive contamination of the litter and the upper layer of the soil would be correspondingly increased. The cleaning of the crowns in coniferous forests would occur three to four times more slowly.

Radionuclides become firmly fixed in the litter and soil, thus fomenting prolonged radioactive contamination of forest communities. From the soil they then move through the roots into the trees, thus increasing the dose accumulated by the internal organs. But all these processes lead to chronic irradiation, the impact of which we will consider below.

Damage to coniferous forests by a year dose does not depend on the time at which the nuclear war begins, since most of the radioactivity would be concentrated in the crowns of the trees in the course of a year. The picture is different for deciduous forests: if radioactive contamination occured in the winter, it would not be absorbed by the crowns of trees. Therefore damage to deciduous forests would be less in winter than in summer. Here we must further take into account the greater radioresistance of trees in a state of physiological rest.

The "nuclear winter" factor would not have a substantial effect on the degree of radiation damage, since the main portion of the dose would be accumulated during the first few days and weeks, because of the rapid decay of a fresh mixture of products of nuclear fission. Therefore, in our estimation of the impact of acute-chronic irradiation we need not take into account the decrease in dose caused by the vertical migration of radionuclides, since radionuclides leave the crowns of the trees relatively slowly.

Let us assume that a nuclear war were to begin during the spring or summer, i.e., during a stage of vegetation. Then the effect of radiation damage as a function of dose can be described in the form of the following Table 3a.

Table 3a. Damage of coniferous and deciduous forest stands as a function of the absorbed yearly dose by tree crowns [66a]

Absorbed dose 10^3 rad	Coniferous	Deciduous
1	Suppression of young growth	No visible effects
1–6	Drying of part of stand, death of young growth, inviability of seeds	Decrease in growth
6–8	Total death of stand	Considerable decrease in growth
8–30	Total death of stand	Drying of part of stand, death of young growth, inviability of seeds
30–50	Total death of stand	Total death of stand

Of course, for mixed forests we have a moderate situation.

It must be noted that the effect of radiation damage to trees does not appear immediately. The data contained in Table 4 show how the dynamic of radiation damage from acute-chronic irradiation develops.

Table 4. Doses absorbed by pine tree crowns, which induce double reduction of indices, characterizing radiation damage of tree layer [32]

Indices	Dose, 10^3 rad
Decrease in sprout length (1 year after irradiation)	1
Decrease in the amount of trees having annual sprouts (2-nd year after irradiation)	2.5
Decrease in radial stem wood growth (3-rd year after irradiation)	3.0
Drying (radiation death) of trees (LD_{50}) of trees (4-th year after irradiation)	3–4

2.5. Radiation damage of forest communities — III. Chronic irradiation

Chronic irradiation (the term of its activity is defined in decades) is mainly determined by contamination from the destruction of atomic power plants and is formed by long-lived isotopes, of which the most dangerous biologically are the following: ^{90}Sr ($T_{1/2}$ = 28 years), ^{137}Cs ($T_{1/2}$ = 30 years) and ^{14}C ($T_{1/2}$ = 5570 years). These isotopes are involved in the biogenic cycles, and accumulated in the tissue of living organisms.

After the decay of short-lived nuclides (several days) the intensity of dose declines substantially, and the remaining long-lived nuclides migrate from the tree crowns to below the canopy, and are accumulated in the litter and the upper layers of the soil. Then an ecosystem would be subjected to chronic irradiation, continued over the course of several decades. Although the intensity of dose accumulated during a period of chronic irradiation is a hundred times lower in comparison to the initial dose, the total dose accumulated over an extended period would be substantial. Therefore, for chronic irradiation the degree of radiation damage to plants is determined less by the value of the total dose than by its intensity. Under these conditions, the value of the total dose, which brings about certain effects of radiation, turns out to be an order of magnitude higher than for acute or acute-chronic irradiation.

The reasons for this, are as follows:

1. In order for a cell to be destroyed immediately after irradiation, a dose greater than a certain threshold value is required. If the doses are lower than the threshold value, the damage is reversible. A low intensity dose accumulated by the cells of chronically irradiated plants during one cycle of cell-fission will be lower than this threshold. Such cells continue to divide normally.

2. A second reason for the high critical value of the total dose for chronic irradiation is connected with the adaptation of cells to radiation. The preliminary irradiation of cells even in relatively low doses significantly lowers their sensitivity to further irradiation by greater doses.

The influence of irradiation over the course of many years (intensity of chronic radiation is 0.1–5 rad/day) on the growth of trees has already been studied (op. cit. [71]). It was found out that irradiated trees as a rule did not increase in biomass. The irradiation of yews and various species of pine by γ-quants with a dose intensity of 10–20 rads/day kills the trees over the course of three vegetation periods.

Even a dose of relatively low intensity (about 2 rad/day) brought about a significant retardation of growth and the dying-off of most needles.

Since the data on radioactive contamination contained in Chapter 1 are expressed in Ci/m^2, it is necessary to determine the scale coefficients which allow us to connect the density of contamination with dose intensity. The dose intensity on various levels for a standard density of contamination of 1×10^{-3} Ci/m^2 (σ_e) is contained in Table 5.

Table 5. Dose intensity distribution (rad/day) with radioactive contamination density $\sigma_e = 1 \cdot 10^{-3}$ Ci/m^2

Height above the soil level, cm	The period of time after irradiation (90 Sr)	
	2 months	8 years
1000	0.3	0.1
10	2.0	0.8
0 (litter surface)	2.9	1.2
− 2	0.3	1.0
− 6	0.01	0.1

It is assumed that radionuclides are distributed uniformly over the forest litter. The intensity of accumulated dose of β-radiation makes up 90 % of the total intensity (note that it quickly decreases with an increase in height). The intensity of accumulated dose of γ-radiation makes up 10 %

of the total intensity (note that it remains constant at any height). Hence, if radionuclides were completely washed out onto the surface of the litter, the standard contamination of 1×10^{-3} Ci/m^2 would bring about a dose intensity of ≈3 rad/day on this surface. The dose intensity in the upper layer of the forest would be 0.3 rad/day.

However, it is necessary to take into account that the rate of migration of radionuclides from the crowns of trees into the litter is rather low for conifers (characteristic time of 3–4 years). In addition, ^{137}Cs easily penetrates from needles and leaves into the internal organs of the tree. If we take these factors into account, the dose accumulated in growth points of conifers will be higher.

Using this information and the information on the density of contamination (see Chapter 1), it is possible to calculate the intensity of dose of chronic irradiation in territories of different population densities (see Table 6).

From this table we can see that a significant effect of chronic irradiation would be observed only in territories of maximal contamination, and then only for coniferous forests. Therefore, the direct effect of chronic irradiation on forest communities would be rather weak.

However, there exists another effect which is more dangerous in its long-term action. The radioactive contamination of the litter and soil would have a strong impact on the seeds and seedlings of trees, thus retarding or completely cutting off the renewal of vegetation.

Table 6. Radioactive contamination density and distribution of absorbed dose intensity

Population density	USSR			I	EUROPE (without the USSR)			II	NORTH AMERICA			III
(ind./km^2)	1	2	3	4	1	2	3	4	1	2	3	4
150	2.7	8.1	0.81	4.05	4.5	13.5	1.35	6.75	12.8	38.4	3.84	19.2
30	0.5	1.5	0.15	0.75	0.9	2.7	0.27	1.35	2.6	7.8	0.78	3.9
10	0.25	0.75	0.075	0.375	0.3	0.9	0.09	0.45	0.9	2.7	0.27	1.35

1 – density of radioactive contamination (10^{-3} Ci/m^2)

2 – absorbed dose intensity of the surface (rad/day)

3 – absorbed dose intensity in the upper layer of forest (rad/day)

4 – absorbed dose intensity in the upper forest layer including retention in forest crowns (rad/day)

2.6. The effect of irradiation on the germinative capacity of seeds and the growth of seedlings

In considering this question it is necessary to take into account that most seeds would be irradiated by significant doses in the course of a year after the nuclear conflict.

According to some data (op. cit. [71]), the loss of the capacity for forming normal seeds can be caused even by sublethal doses (30–40 % of LD_{100}). For example, the pollen of Pinus strobus and Pinus rigida completely lost its viability under doses of 11 rad/day and above. Under doses of 7 rad/day, 50 % of the pollen was under-developed (at the moment of maturation), and under a dose of 3–5 rad/day, the germination of pollen grains was two times lower. In deciduous species the generative organs are more stable under irradiation than in conifers. Thus for various species of oak the germinative capacity under a dose of 4–6 rad/day was decreased by 25–40 %. But under a dose of 25 rad/day almost all seeds were non-viable. There is yet another aspect to this problem. Irradiation changes the morphology of the flower, retards flowering and fertilization. For example, a 5-day retardation in flowering in the oak was observed under a dose of 20 rad/day, and this retardation can have serious consequences, since the female sexual cells of the oak are capable of fertilization for not more than 5 days.

Critical doses of acute irradiation of the seeds of conifers (the reduction of germination capacity by 50 % and the destruction of most of the seedlings) are from 600 to 6000 R of acute irradiation; for the seeds of deciduous trees, these doses are 10,000 R and above.

What happens when seeds which have maintained their viability begin to germinate in soil contaminated by radionuclides?

The most critical period in the life-span of the seedling is the initial one.

During this period practically all important tissues of the seedling are subjected to irradiation. The complete loss of ground germination in pines takes place under a dose greater than 2500 rads. But suppression begins to occur under a dose of about 1000 rad [71]. A significant suppression of growth and of the development of seedlings is observed at levels of radioactive contamination of the soil above 2×10^{-3} Ci/m². The intensity of accumulated dose on the level of the terminal bud of one year old seedlings is more than 1 rad/day. The suppression takes the form of a reduction of seedling growth in height or the cessation of growth. So the number of seedlings with dried up terminal buds at the end of the third vegetation period is 30–40 % for a contamination of $2.6–5 \times 10^{-3}$ Ci/m². For a contamination of 5×10^{-3} Ci/m², a dying off of the root systems in the forest litter is observed. For a dose intensity of 4 rad/day, it does not bring about irreversible changes, although the irradiation reduces the growth of seedlings. However, the seedlings which grow out of irradiated seeds are subject to fungus disease much more often. The dose intensity under which damage to seedlings is irreversible is 10–15 rad/day, which corresponds to a contamination density of $6–9 \times 10^{-3}$ Ci/m² [71].

The renewal of a damaged forest is possible not only through seeds, but also by vegetative means. However, the capability for vegetative renewal is found mainly in deciduous species, and this capability is conserved even after irradiation by large doses (30,000–50,000 rad). If we take the above into account, we can conclude that after a nuclear war deciduous forests can be renewed (if only by vegetative means). The renewal of coniferous forests is rather problematic, since this happens mainly through seeds in the presence of high-quality seed material of local origin.

2.7. Radiation damage of other ecosystems

For the estimation of radiation damage of other ecosystems we can use the empirical generalization that the degree of radiation damage, determined as a function of the relative dose (% of LD_{100}), is identical for many species of plants (see Table 7). Species specificity of radiosensitivity is manifested only in differing values of the lethal dose LD_{100} [71].

Using the data contained in this table and in Table 1, which contains data on the degree of damage as a function of dose for various types of communities, it is possible to determine the degree of damage of various communities for a given dose. For example, the dose causing sterility of pollen in meadow grasses is 40,000 to 60,000 rad (LD_{100} = 100,000–150,000 rad).

Table 7. Relative doses causing various radiation effects in plants (with chronic γ-irradiation)

Reaction	Number of species	Doses in % of LD_{100}
No visible effects of radiation damage	14	11
Decrease in growth rate by 10 %	23	26 ± 2,5
Inviability of seeds	8	31 ± 5,5
Decrease in growth rate by 50 %	12	34 ± 3,5
Pollen sterility	4	41 ± 4,5
Considerable inhibition of growth	41	58 ± 3
LD_{50}	17	75 ± 2,5

2.8. The impact of irradiation on animal populations

Among animals, insects are the most radioresistant ones, but their radiosensitivity depends essentially on their phase of development. The embryonic stage is most sensitive (LD_{100} = 100–1000 R for acute γ-radiation). During the larval stage the radioresistance is increased (LD_{100} = 4000–6000 R). The radioresistance of the pupal and imago stages is much higher. For example, a 30-hour old pupa of drosophila can withstand a dose of 12,000 R, and at 50 hours a dose of 80,000 R. LD_{100} for the imago of drosophila is 100,000 R.

Mammals and birds are least radioresistant. Acute irradiation in doses of several hundred Roentgen are lethal for adult organisms of most species. Even a dose of 25 R has a negative effect on the prenatal development of mammals [71]. Table 8 gives a fairly graphic sketch on the radiosensitivity of various species (op. cit. [71]).

Table 8. Irradiation doses lethal for some mammalian and bird species ($LD_{50/30}$ denotes a dose that brings about the death of 50 % of irradiated population in the course of 30 days

SPECIES	$LD_{50/30}$, R
Small rodents	300–600
Goat	350
Raccoon	580
Lynx	580
Gray fox	710
Rabbit	750–820
Linnet	400
Bramble finch	500
Goldfinch	600
Greenfinch	600
House sparrow	625
Pigeon	920

In estimating the death of animals it is necessary to take into account their ecological niche. For example, birds living in the crowns of trees receive a greater dose of β-radiation, since the radioactive fallout in the first few weeks would be absorbed mainly by the crowns. Since these doses would be thousands of rads, all of these species would die out.

Large terrestrial mammals would receive a much smaller dose (mainly γ-radiation). The chronic dose of radioactive contamination falling on the surface of the earth would be significantly lower for large mammals than for small ones, since the path length of β-particles in the air is relatively small (about 1 m). In addition, most β-radiation would be absorbed by external organs. A much more significant effect of chronic irradiation is the accumulation of internal dose from consuming contaminated vegetation.

Rodents will receive a relatively small dose, since most β-radiation is absorbed by the layer of soil above their burrows, and they spend little time on the surface. For instance, the irradiation of a population of mice over a 15 day period (with a surface dose of 5690 rads) led to a reduction of population size of only 38 %, although LD_{100} = 500 rad for this species.

In conclusion let us consider the possible mechanisms of the impact of surface contamination on man. The accumulated dose will be mainly determined by γ-radiation. However, taking into account the fact that there is always a layer of dust above the surface of the soil, grass also partially catches radioactive dust, and gonades are located on the average at a level of 1 m above the surface, the dose accumulated by the gonades would be approximately 2–3 times greater than the γ-dose. The same can be said about the gonads of large mammals. Consequently, in estimating the impact of radioactive contamination on reproductive capacity, it is also necessary to take into account β-radiation.

2.9. Secondary effects of radiation damage to ecosystems

In a climax ecosystem each species occupies a certain ecological niche. The elimination of any species disturbs the equilibrium of the ecosystem; more radioresistant species would begin to occupy its ecological niche. In addition, each species in an ecosystem is one link in the food chain. The elimination of one link invokes a change in the entire chain: the death of producers automatically leads to the death of consumers; the death of "predators" disturbs the control of "prey" populations, etc.

The death of trees leads to the beginning of an active growth of dwarf shrubs and grasses, perennials capable of vegetative renewal.

In a forest damaged by radiation, conditions are favourable for massive outburst of insect-pests. This effect is heightened by the death of birds, which control the population of insects. For example, in a study of pine trees weakened by irradiation, it was shown that 15–20 % were infested by pine beetles (5–7 % for non-irradiated trees) [32]. In an irradiated forest, for example, a 200-fold increase in the population size of leaf-eating insects was observed. The result may be that the damage from insect-pests is greater than from the direct impact of ionizing radiation.

Yet another secondary effect is the reduction of the decomposition rate of litter (which leads to a reduction in the rate of biological cycles of matter) on account of the relatively low radioresistance of soil fungi and earthworms. For example, the total population of earthworms in the soil of an irradiated forest (for a dose of 7000–20,000 rad) was 10 times lower, and the biomass of soil fungi was reduced by half.

2.10. The geographical distribution of radiation damage of communities

Given the geographical distribution of various plant communities (more exactly biomes) and the geographical distribution of doses and radioactive contamination, and using the information on the dependence of the degree of damage on dose, we can estimate (by superimposing two maps) the geographical distribution of radiation damage to various communities.

A map of the distribution of vegetation (various biomes) in the regions under concern is depicted in Figures 6a and b.

The maps of radiation damage to vegetation depicted in Figures 7a and b were created as follows. The maps of the distribution of doses and contamination (Figures 3a, b) were superimposed on the maps of vegetation (Figures 6a, b), thus giving us the distribution of doses and contamination in territories with various types of vegetation communities. Then, using mainly Table 1, as well as other information on the dependence of the degree of radiation damage on the accumulated dose and intensity of radioactive contamination, we get a scale of high, moderate and low degrees of radiation damage to various vegetation communities.

a) Forest communities

1. *Low degree of damage:* reduction in needle growth, increase in the falling off of old needles, reduction of shoot length, suppression of the young growth; for deciduous trees – the partial drying up of leaves.

2. *Moderate degree of damage:* reduction in growth by 10–30 %, decrease in the number of trees bearing annual shoots, inviability of seeds, the death of most young trees.

3. *High degree of damage:* significant reduction in growth rate, sterility of pollen, retardation of generative organ formation, death of all young trees, drying up of the crowns, and death of all trees (it is possible for deciduous trees to be renewed vegetatively).

b) Grassland communities

1. *Low degree of damage:* slight suppression of growth, decrease in the number of shoots, slight reduction in the diversity of species.

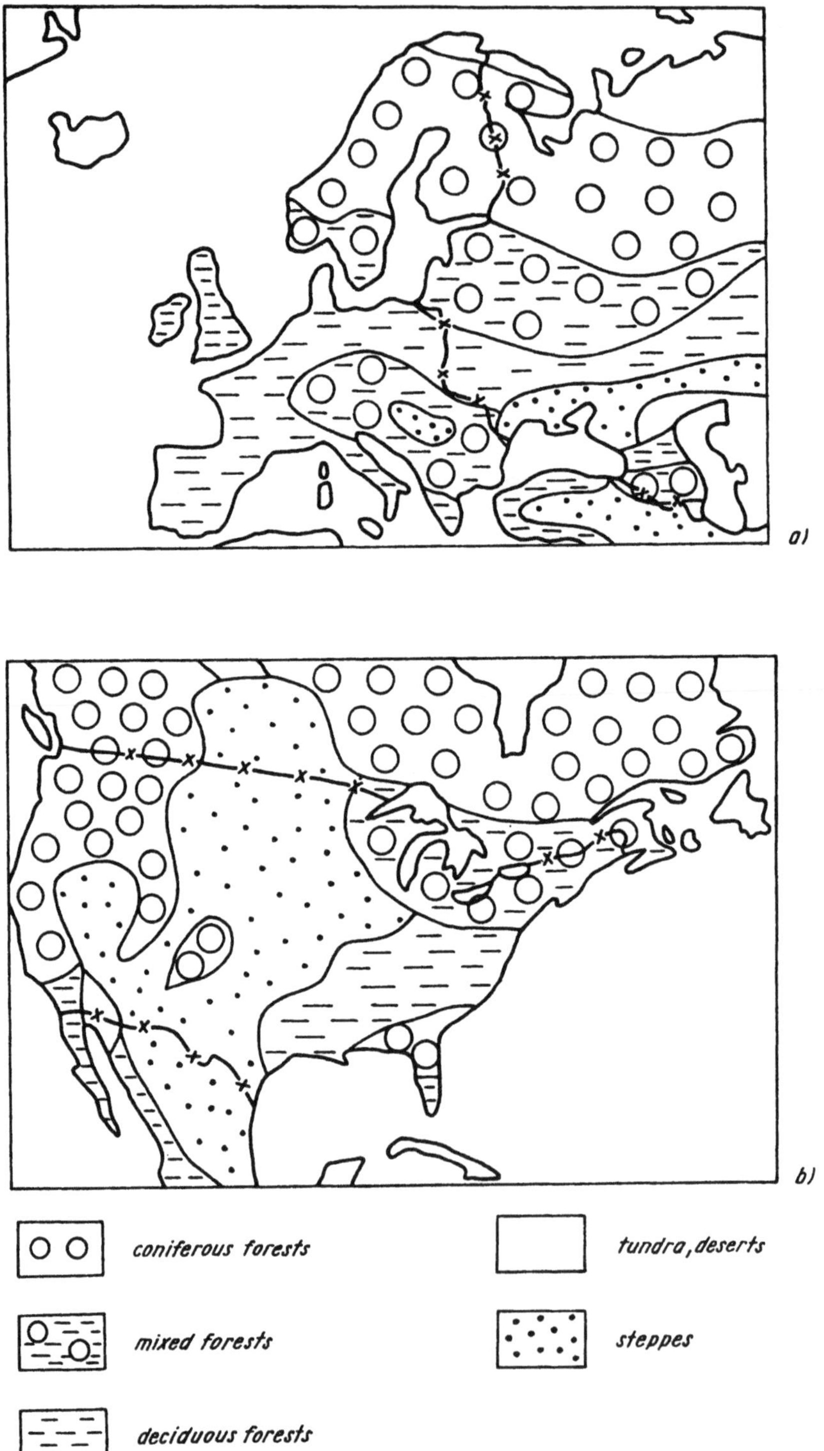

Fig. 6. Distribution of various types of plant communities in regions of Europe (a) and North America (b)

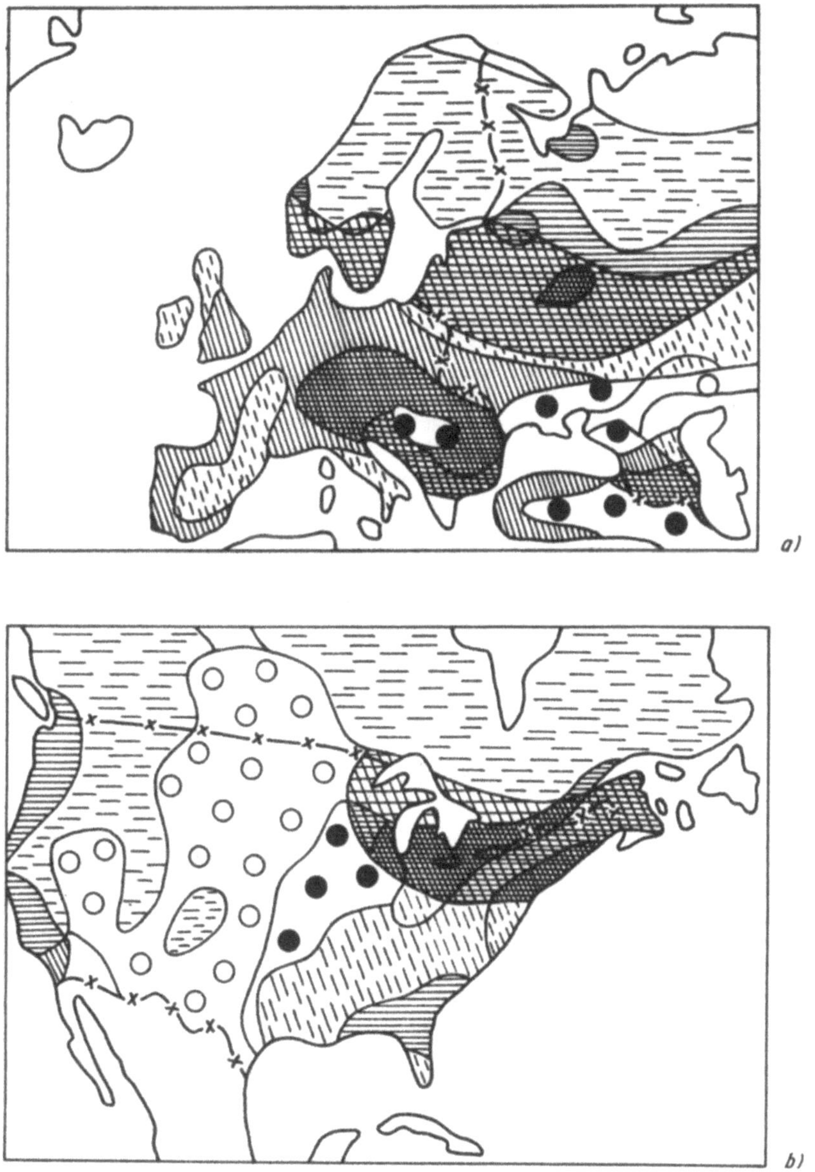

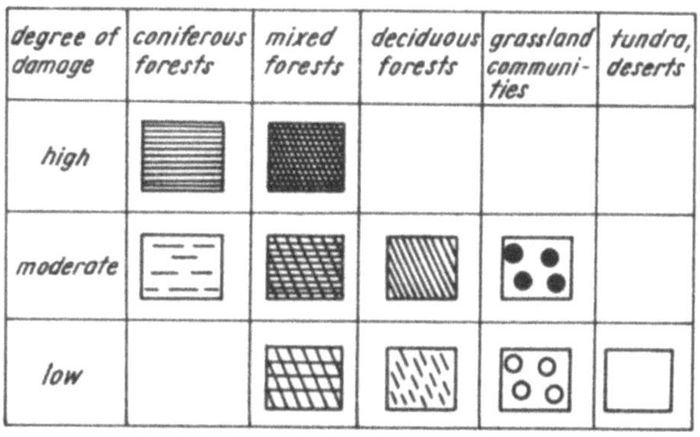

Fig. 7.
Damages of vegetation in regions of Europe (a) and North America (b) one year after the beginning of nuclear war. Effects of "radiation shock" and "long-living" radioactive contamination were taken into account. Effects of "nuclear winter" and forest fires were not considered

2. *Moderate degree of damage:* significant suppression of growth, partial sterility of pollen and inviability of seeds, death of some dwarf graminoids, shrubs, and decrease in the renewability of annuals, significant decrease in the diversity of species (shift towards sedges).

Since in areas suffering the greatest damage the density of radioactive contamination by long-lived isotopes is greatest (from 3 to 12 mCi/m^2), renewal of coniferous forests in these territories would be practically non-existent. Even the remaining forests would be doomed to elimination. In territories suffering less damage, the density of radioactive contamination ranges from 0.55 to 2.6 mCi/m^2, which is sufficient to cause a significant reduction in the growth rate of conifers. Renewal in conifers, if possible at all, would be very slow. If we take into account the death of most young trees and the inviability of seeds (as a result of acute and acute-chronic irradiation) and the synergistic effect of other factors in a nuclear war, the very existence of a coniferous forest biome becomes problematic.

The remaining coniferous forests would be in a state of suppression and would be easy targets for pests. When seen against the background of weakened forests, the impact of other unfavorable factors would be further strengthened. Territories where a forest died out would experience the spreading of radioresistant species of plants: grasses, shrubs, ferns, mosses and lichens. The type of ecosystem would be changed — the so-called "nuclear succession". The ecosystem with a high productivity (forest) would be replaced by an ecosystem of low productivity (grassland).

As far as animals are concerned, we can see in Figures 3a, b that in territories with a high population density doses of acute γ-radiation are sufficient to bring about the death of most species of mammals and birds. The doses of β-radiation in all three regions are sufficient to bring about the death of birds living in the crowns of trees and in the vegetation cover of grassland communities. Individuals that do not receive lethal doses become partially or completely sterile, which also leads to the gradual extinction of these populations.

2.11. Conclusion

The "radiation shock" leads to serious radiation damage or radiation death of coniferous forests over most of Europe and North America. Mixed forests would also be seriously damaged. Other biomes would suffer a much lower degree of damage.

The inviability of seeds, sterility of pollen, damage to young trees as a result of the "radiation shock", and the impact of chronic irradiation on seedlings lead to a state in which coniferous forests in the zones of greatest contamination would not renew. In zones of moderate damage, very slow renewal of conifers would be possible, though it is of low probability if we take into account the synergistic effect of other factors of nuclear war. The vegetative renewal of deciduous forests is possible.

As a result of the "radiation shock" many species of mammals and birds would die out. Among the survivors many individuals would be sterile (because of chronic irradiation by long-lived isotopes). This greatly increases the probability of the elimination of remaining populations.

In conclusion we note that, using the information contained in this chapter and calculations of dose distribution and radioactive contamination by region contained in the previous chapter, it is easy to make more localized and more detailed estimations of the degree of radiation damage to various ecosystems.

3. BURNING OF VEGETATION (FOREST AND GRASSLAND FIRES)

3.1. Introduction

Large-scale forest and grass fires would be one of the main factors of a nuclear war having an impact on the biota. Agricultural lands would also be burned, but forest fires would still play the major role. This is connected, first of all, with the fact that in the USA and especially in Canada and the USSR, forests are located in the direct vicinity of strategic centers, and "civilization fires" trigger the mechanisms of forest fires. Secondly, the amount of combustible organic material in forest communities is an order of magnitude higher than the amount of such material in grassland communities. This means that forest fires make the greatest contribution (in comparison with fires in other ecosystems) to changes in the chemical make up of the atmosphere and to an increase in the level of pollution after a nuclear war.

Although large-scale grassland fires present a situation fairly common in the biosphere (in an average year up to 60 % of grasslands burn [12]), analogues of "nuclear" forest fires are not to be found in history. Therefore our main method of estimation will be extrapolation by analogy. For this it is necessary to analyze the situation of "ordinary" forest fires.

3.2. Some characteristics of "ordinary" forest fires

On the average 7.2×10^4 km^2 of forest area in the northern hemisphere burns annually [61], but in some years fires spread over a much greater territory. A gigantic fire in Siberia in 1915, in which 1.4×10^6 km^2 were burned, is described in [75]. However, the other authors [44] consider this estimate to be exaggerated. Ordinarily the scale of forest fires (FF) is much smaller. In Gascony (France), in 1949, 1300 km^2 of forest was burned; in 1961, in the southwestern USA, 240 km^2 of forest was burned, etc. [66].

According to the classification in [49], FF can be divided into lower, upper, and underground (peatbog) ones.

Lower FF, in which forest litter, lichens, mosses, grasses, and fallen branches burn, are the most widespread. Upper FF affect trees, the grass-moss cover of the soil, and seedlings. In underground (peatbog) FF, the peat layer and roots of trees burn. FF can be of a combined character. Lower FF occur mainly in the spring; all types occur in the summer and autumn; FF rarely occur in winter. However, peatbog fires sometimes spread under snow.

The area of lower FF makes up 76–84 % of the total burned area; the area of upper FF – 16–24 %; the area of underground fires is usually not more than 0.1 % [12]. In dry years this relationship can change. In 1938, in the USSR the area of upper FF was 51 % and the area of underground fires in 1939 rose to 6 % [55].

What can we say about the causes of forest fires? In the forests of northern Europe, approximately 97 % of all FF are of an anthropogenic nature, while only 3 % are started by natural causes (lightening bolts) [12]. Anthropogenic FF are distributed over a territory extremely non-uniformly: up to 60 % of all fires occur within a 5 km zone around populated areas; up to 90 % occur within a 10 km zone [43].

The appearance of FF is closely connected with weather conditions. The most dangerous period is summer [49].

The distribution of FF by month is different for each region. An example of the average distribution of FF by month in the forests of northern Europe is contained in Table 1 [12].

Table 1. Average distribution of FF in the forests of northern Europe

month	percentage of FF
May	15
June	31
July	35
August	17
September	2

Litter-humus and peatbog FF, in contrast to above-soil FF, which usually begin in May, occur most often in the second half of the summer, when the moisture of the litter and peat is at a minimum. If it is a dry year, peatbog FF continue throughout the winter and last for many years.

FF are an ecological factor which has a many-sided influence on forest ecosystems: on the structure and species composition of stands of trees, renewal, hydrothermal conditions, biogeochemical cycles in the soil, soil nutrition of trees, on trophic links and the dynamic of the reproduction of heterotrophic organisms for various levels of organization [16].

FF significantly change the regime of mineral nutrition in stands of trees because of the burning of litter and humus, and the effect of high temperatures on microbiocoenosis [4, 59]. Depending on the type of soil and its water regime, which influences the leaching rate of ash elements during the period after the fire, the amount of phosphorous, potassium, magnesium and in some cases nitrogen in the soil may be increased [59]. Soils of light mechanical composition may become poor in biogenic elements [18]. The damage caused to forests by FF is not limited to the burning of trees directly by fire. By weakening the trees they may trigger large-scale outbursts of pests [16, 30]. The danger of pest outbursts in burned-out areas further consists of the fact that insects, which had previously developed in trees damaged by fire up to this time, now settle in trees that are potentially stable and capable of resuming their normal life functions [30]. (Note that this is in complete analogy to the secondary effects in forests damaged by radiation, see Chapter 2).

The type of succession following the fires depends on many factors. The replacement of coniferous species by deciduous species is possible, but does not always occur [12]. In the taiga forests of the north and northwestern USSR, pallustrification of burned-out areas is a common phenomenon [69]. This is due to the fact that between the Arctic Circle and a latitude of 60° N, there are two subzones of taiga forest: the northern taiga subzone of glay-podzol soils, and the middle taiga zone of podzol soils. Gley-podzol soils are characterized by clearcut gleyification of the upper part of the soil profile and by formation of peat litter.

A high level of ground-water (40–60 cm) is characteristic for this type of soil because of the prevalence of two-component soil rocks – sands and loam sands – bedded on loam or clay at a depth of 40–60 cm. A rise in the level of ground-water to a level above 40 cm leads to the rapid development of processes of pallustrification. A rise in the level of ground-water can be caused by cutting or by fire. The destruction of trees, which actively transport soil water and catch up to 30 % of rainfall, leads to an increase in input water balance of approximately 200 mm [36]. Since the average porosity of the ground is 50 %, the level of ground-water may rise to 40 cm. However, actually the rise in the level of ground-water will not be this great.

Since the vertical movement of liquids is bounded by the waterproof layer, and horizontal movement is bounded by the conditions of lack of flow through the boundary of the area considered, the amount of rise in the level of ground-water depends on the size of the burned-out area. If one fifth of the forest is destroyed, the level of ground-water will rise 8 cm. If we take into account the fact that the average level for gley-podzol soils is 50 cm, and if we consider the symmetrical distribution, we can conclude that if, for example, 20 % of the forests in these territories are burned, the remaining area will be reduced by an additional 40 % due to pallustrification.

As far as fauna is concerned, species of forest birds usually disappear after a fire. The wide spreading of fires leads to the expansion of the European type of fauna and at the same time restricts the role of Siberian elements [41].

In other regions the successions after a fire may proceed differently. For example, in the Far East burning out leads to the expansion of Manchurian meadow-forest forms and arctboreal elements, accompanied by a strong depression of nemoral, ancient-taiga and tropical forest forms [49].

In contrast to the USSR, the pallustrification of burned-out areas in North America is possible only in the area surrounding Hudson Bay because of differences in soil types [107].

3.3. Estimation of the scale of FF after a nuclear war

According to [98], self-combustion of a forest occurs at points where the heat impulse from a nuclear explosion exceeds 15 cal/cm^2. The amount of burned organic mass in the core of combustion is almost independent of season, weather, character of vegetation, and depends only on its mass.

The average forestation in the northern hemisphere is 35 %.

After the explosion of a nuclear bomb of 1 Mt, 500 km^2 of forest would be burned [98]. Another estimate: an explosion of 0.4 Mt would burn out about 250 km^2 [86]. If we use these estimations to calculate the effects of fires for a total yield of 5400 Mt, taking into account the percentage of forestation, we get the total area of burned-out forests: $0.945–1.18 \times 10^6$ km^2 $\approx 10^6$ km^2. This value represents the direct effect of nuclear attacks. An area of 10^6 km^2 is 4–5 % of all forests in the northern and central latitudes.

The above estimations of burned-out area are applicable for most of the year; however, the probability of the appearance of widespreading fires initiated by "nuclear" fire is raised during the summer months [113].

The likelihood of further spreading of the fire after ignition depends on the presence of combustible material (vegetation), its distribution throughout the territory, and its state (moisture content).

The main role in the spreading of FF is played by the layer of major conductors of combustion in the soil (mosses, lichens, and fine particles of vegetation).

The main conductors of combustion are rather hygroscopic; they dry up easily, but they also easily take on moisture from rainfall and high humidity. They cannot burn under snow. The term of fire danger in a year is as follows: for a latitude of 50° – 185 days, 55° – 160 days, 60° – 130 days, 65° – 100 days, 70° – 80 days [49]. During the fire season there are rainy periods when FF practically do not spread. In southern taiga forests, fires do not spread during the summer, when green grasses are prevalent and when deadfall is actively decomposing in deciduous forests. These days of low danger make up 30–60 % of the fire season. Therefore, the number of days when FF may spread in a forest zone is approximately 20 % of the year. The usual propagation rate of FF is 100–300 m/day, but in open areas covered by dry grass it is an order of magnitude higher (for mean wind velocity). This is the situation for "normal" fires. It is quite possible that if a huge area of forest is ignited at once, the FF may enter a stage of *self-development,* i.e. it will turn into a "patchy" fire of the explosive type or into a "fire storm". Under natural conditions such fires develop, as a rule, during the afternoon in the presence of a strong wind. During the night and especially in the morning hours, when the wind is weak and the conductors of combustion have a greater moisture content, the development of "patchy" fires is unlikely. However, after a nuclear explosion much greater areas (in comparison with the directly burned-out area of the forest) would be "dried up", the dependence on the time of day would be weakened, and the probability of the appearance of "patchy" fires would rise sharply.

The spreading of fire under natural conditions is hindered by swamps, rivers, lakes, etc., or by rainfall. Various artificial barriers (fire alleys, roads, etc.) are overcome fairly easily by large-scale fires. An analysis of the information on FF shows that there are limits to the spreading of "patchy" fires which are independent from its causes. The typical area of burn-out does not exceed 800–900 km^2.

The number of targets according to the AMBIO scenario is 14744, with a total nuclear yield of 5742 Mt. Since for the regions being considered the total nuclear yield is 5400 Mt, if we assume the yield and number of targets to be proportional, we get a total number of targets in these regions of 13866. If we take into account the fact that forestation is 35 %, the number of targets in forested territories is 4853. If an attack on each target initiates a widespreading fire, burning on the average on 850 km^2, we can easily calculate the area burned directly after a nuclear war to be 4.125×10^6 km^2 or approximately 4 million km^2.

Since the term of fire danger is approximately 20 % of the year, there is an 80 % chance that non-spreading fires will arise after nuclear attacks (area of burned forest – 10^6 km^2), and a 20 % chance that spreading fires will arise (area of burned forest – 4×10^6 km^2).

Since it is impossible to foresee the timing of the nuclear war, and therefore to evaluate the probability of the appearance of spreading fires, in the calculations that follow we will use two estimations: the minimal (corresponding to a burned-out area of 1×10^6 km^2) and the maximal (corresponding to a burned-out area of 4×10^6 km^2) one (Table 2).

Table 2. The burned-out area of forests in various regions

Region	USSR I	Europe II	North America III
area (10^6 km^2)	5.6	4.7	10.0
forestation (%)	32	29	32
forest area (10^6 km^2)	1.8	1.4	3.2
burned-out area (10^6 km^2)	0.3–1.3	0.3–1.3	0.3–1.3
% of burned forest	19–75	23–92	10–41

3.4. The composition and quantitative value of products produced by FF

According to [82], forests with an area of 1–4 mil. km^2 contain approximately (2.2–8.8) X 10^{10} tons (t) of dry material. In a typical FF approximately 25 % of all organic material burned [113]. Since the dry matter of forest ecosystems is composed of about 40 % carbon, 0.4 % nitrogen and 0.06 % sulphur [8], the products of combustion would be composed of approximately (2.2–8.8) X 10^9 t carbon, (2.2–8.8) X 10^7 t nitrogen and (3.3–13.2) X 10^6 t sulphur.

In a FF 75 kg of aerosols are produced for every ton of burned dry material [86]; thus a fire with an area of (1–4) X 10^6 km^2 would produce (4.1–16.5) X 10^8 t of aerosols. This study [86] also contains another estimate obtained from the results of an analysis of FF in Alaska, namely 35 kg per ton. If we use this estimate, the value (4.1–16.5) X 10^8 t must be reduced by half.

Finally, it can be determined that as a result of FF arising after nuclear attacks, approximately (0.25–1) X 10^{10} of carbon, (0.25–1) X 10^8,t of nitrogen and (0.3–1.3) X 10^7 t of sulphur will be emitted. (0.2–1.6) X 10^9 t of aerosols will be emitted, a figure two times greater than the amount of dust of anthropogenic origins emitted yearly [86].

If we now compare these estimations (using the maximal estimation) with estimations of the emissions caused by urban fires – "fires of civilization" (see Chapter 5), we can see that FF and urban fires emit an approximately equal amount of carbon (10^{10} t), but "fires of civilization emit considerably more nitrogen and sulphur (10^8 t and 7 X 10^8 t of nitrogen and 1.3 X 10^7 t and 3 X 10^8 t of sulphur for forest and urban fires, respectively).

3.5. The influence of FF on the optical properties and chemical composition of the lower layer of the atmosphere

The weakening of solar radiation as a result of large-scale FF has already been observed many times (even at a great distance from the site of the fire). This phenomenon during gigantic fires in the USA (Minnesota and Wisconsin) in October 1918 and in Canada in 1950 is described in [86]. As is shown in this investigation, if a FF were to continue over the course of two months and were to burn 10^6 km^2 of forest, the aerosol emission of (2–4) X 10^8 t over half of the northern hemisphere would create serious dust pollution. As a result, the average solar radiation reaching the earth's surface during the summer at noon would be decreased 150 times (maximal estimation). The decrease in crop yield during a year of large-scale FF has been observed. For example, in 1915, when a gigantic FF took place in Siberia, the vegetation period for all agricultural crops was extended by 2–4 weeks [75]. This was due to both a decrease in PhAR (photoactive radiation) because of a darkening of the atmosphere by smoke and soot, and a decrease in the sum of active temperatures because of a fall in temperature.

Let us now consider the change in the chemical composition of the lower layer of the atmosphere as a result of large-scale FF, when 10^{10} t of carbon, 10^8 t of nitrogen and 1.3 X 10^7 t of sulphur would be emitted (maximal estimation).

Carbon is mainly emitted in the form of carbon dioxide (CO_2). This does not have a significant influence on the concentration of CO_2 in the atmosphere (an increase of 1.5 %). What is important is that approximately 15 % of the carbon is emitted in the form of carbon monoxide (CO) [85]. The CO in the atmosphere is thus increased 7 times. If we then take into account the emissions of carbon from "fires of civilization", the situation would become catastrophic (see Section 5.4.). In this chapter the maximal estimation of emissions from FF is used. Other pyrotoxins would be produced along with carbon monoxide: tens of thousands of tons of ethylene (C_2H_4) and propylene (C_3H_6), the major components of photochemical smog. The concentrations of peroxiacetylnitrates (PAN) and ozone (O_3) would considerably increase. Ozone increases the intensity of leaf respiration, which leads to the death of plants from exhaustion; PAN blocks Hill's reaction in photosynthesis and the plant dies from lack of nutrients [118]. The increase in the concentration of ozone during large-scale FF has been observed by many investigators [91].

Nitrogen and sulphur are emitted into the atmosphere in the form of oxides (NO_x and SO_2). NO_2, for example, is toxic for plants even in small concentrations. However, the main effect would be the following: since a large amount of aerosols would be emitted into the atmosphere, oxides of nitrogen and sulphur would be concentrated in aerosol particles, come into contact with droplets of water, turn into nitric and sulphuric acids and fall to the earth in the form of "acid rain" (see Section 5.3.).

In addition, we should also point out the synergistic interaction of ozone and sulphur dioxide (SO_2), which have a stronger impact when being mixed [88].

3.6. Secondary "post-huclear" and peatbog fires

After the first wave of fire has passed, secondary fires in ecosystems that are destroyed or weakened by the effects of nuclear war (fires, radioactive and "ordinary" pollution, etc.) would occur frequently in the years following a nuclear war.

It is unlikely that fire-control services would be maintained after a war, but at the same time the probability of outbreaks of fires would be somewhat decreased, since ecosystems in the most fire-susceptible zones surrounding cities and industrial areas would already have been destroyed by primary fires (and these structures themselves would be unlikely to remain intact).

Currently the greatest percentage of fires is caused by carelessness. On the one hand, the population after a nuclear war would decrease sharply, but on the other hand, it cannot be assumed that fire-prevention measures would continue to be observed. Therefore we will hypothesize that the probability of outbreaks of fires in the forest would be the same as it is now. The weakening of the forest, more specifically the increase in the percentage of dead wood, would lead to the extension of the period of fire danger, as well as to an increase in the area burned. However, it is difficult to make any kind of estimations. Since 0.3 % of all forests in the northern hemisphere burn every year (7.2×10^4 km^2) [61], we can conclude that an additional 0.3 % of the remaining forests would burn as a result of secondary fires.

After nuclear attacks peatbog fires may occur. Under normal conditions the percentage of these fires is fairly small – 0.1 % [12], but in dry years it may be many times greater (6 % according to the estimation given in [55]). Let us assume that all peatbogs in the regions of a nuclear conflict would begin to burn as a result of nuclear attacks.

The total volume of peat in the countries of the northern hemisphere is about 2×10^{11} t (at 40 % humidity) [73]. Peat contains 50–60 % carbon, 1–3 % nitrogen and 0.1–1.5 % sulphur [46]. The total area of forests in the northern hemisphere is 24–26 mil. km^2. If fires covered an area of $1–4 \times 10^6$ km^2 and all peat in this area were burned, only about $(0.5–2) \times 10^{10}$ t of peat (dry mass) would be burned. As a result, approximately $(0.25–1) \times 10^{10}$ t of carbon, $(1–4) \times 10^8$ t of nitrogen and $(0.4–1.6) \times 10^8$ t of sulphur (given an average composition of 55 % C, 2 % N, and 0.8 % S) would be emitted into the atmosphere. If we compare these emissions with the emissions from FF (see Table 3), we can see that peatbog fires cause more pollution of the atmosphere (especially nitrogen and sulphur), than FF.

However, the principal difference between them lies in the rate of combustion, and consequently in the *intensity* of emission. The rate of peatbog fires is several orders lower than the rate of FF (according to various estimations, peatbog fires can last from 2 to 3 years). Therefore, while FF bring about massive "shock-emission" of pollution into the atmosphere, peatbog fires cause a continuous flow of pollution over the course of many years. Since we are more interested in the effects of "shock" pollution, we will not consider this flow in the calculations that follow (Chapter 5). However, it should be noted that peatbog fires can have a substantial influence on the global biogeochemical cycles of nitrogen and sulphur.

Table 3. Emissions of carbon, nitrogen and sulphur for forest fires and peatbog fires (in millions of tons)

Type of fire	C	N	S
Forest fire	2200–8800	22–88	3.3–13.2
Peatbog	2750–11,000	100–400	40–160

3.7. Conclusion

The effect of large-scale fires that arise after a nuclear war and that lead to the burning out of large areas of forest can be of a global character.

From 1 to 4 million km^2 of forest would burn immediately after nuclear attacks. As a result, 19–70 % of the forests in the European part of the USSR (including the Urals and Caucasus), 23–92 % of the forests in Europe (excluding the USSR), and 10–40 % of the forests in the USA and southern Canada would be burned.

In northern Europe (60–65° N) this would lead to the progressive pallustrification of more than half of the remaining taiga forest. In North America this effect would be weaker – only the territory surrounding Hudson Bay would be affected.

As a result of the burning of forests, huge amounts of carbon (in the form of CO_2 and CO), oxides of nitrogen and sulphur would be emitted into the atmosphere, thus substantially altering its chemical composition. The pyrotoxins produced by FF would have a significant toxic impact on the biota.

Peatbog fires, which would last for years and would emit a significant amount of nitrogen and sulphur, could have a substantial impact on the global biogeochemical cycles of these elements.

As far as grassland fires and the burning of agricultural areas is concerned, many observations have shown that in these regions a fire, once it appeared, would spread "from river to river". These fires can begin at practically any moment in the snow-free season. Therefore, we can conclude that if a nuclear war begins in the snow-free season (for the territories being considered), almost all grasslands and agricultural areas would be burned. However, the influence of these fires on the total pollution of the atmosphere would be insignificant (in comparison with pollution from forest fires), since the surface biomass of these ecosystems is many times smaller than the biomass of forest ecosystems. In addition, renewal of these ecosystems would occur rather quickly.

The explanations above allow us to conclude that the large-scale burning of vegetation after a nuclear war would have a serious impact on the further evolution of the biosphere.

4. "NUCLEAR WINTER" AND "NUCLEAR NIGHT" AND THEIR IMPACT ON ECOSYSTEMS

4.1. Introduction

The impact of the "nuclear winter" and the "nuclear night" on ecosystems is one of the most difficult factors to estimate. While a factor such as radiation could be evaluated on the basis of nuclear weapon tests, obviously the "nuclear winter" and the "nuclear night" have not been encountered in the history of the biosphere. Therefore, for our analysis we will first consider some general aspects of the impact of low temperatures and low illumination on ecosystems; we will then use this rather poor information on the effect of factors similar to the factors of the "nuclear winter" to arrive with our estimations. (Here and below we will use "nuclear winter" for both the "nuclear winter" and "nuclear night".)

4.2. "Nuclear winter" – decrease in temperature and illumination

In estimating the decrease in temperature and illumination during the "nuclear winter", we used the results of calculations according to a combined model of general atmospheric and oceanic circulation [77]. For more precision the effect of large-scale hydrodynamic transportation of optically active admixtures on the evolution of climatic characteristics was also estimated [67].

The almost total blockage of solar radiation by "nuclear" aerosols would lead to a rapid decline in temperature on the surface of the mainlands of the northern hemisphere. Over a 15-day period the temperature of the lower air layers would fall by 10–50 °C, and then would begin to climb slowly.

The large-scale hydrodynamic transportation of optically active admixtures is especially significant for equatorial territories and for the lower latitudes of the southern hemisphere, it determines the main effect of a drop in temperature. For example, in a scenario with a total yield of nuclear detonations of 10,000 Mt, in a month the temperature in the tropics would fall to 0 °C. In three months the wave of pollution would reach Antarctica. The long-term reduction in temperature in the southern hemisphere would average 5–8 °C, a decrease significantly greater than if calculated without considering transportation of aerosols. The cooling of southern oceans changes the dynamic of the "nuclear winter" and increases the duration of the temperature reduction. The calculated temperature reductions of the layer of air of the earth are contained in Figure 8. These results will then be used to estimate the reactions of land ecosystems to the "nuclear winter". It should be noted that although these data were obtained for a scenario of a total exchange of 10,000 Mt, while we use the AMBIO scenario elsewhere (total exchange of 5742 Mt), the calculations show that there is no substantial difference between these two scenarios in temperature and illumination. The results of our calculations of illumination are contained in Table 1.

Table 1. Decrease in illumination in comparison to normal illumination on the 15-th day of the "nuclear winter". Values of the decrease in illumination are rounded to 0.1 %

Latitude	90N	78N	66N	54N	42N	30N	18N	6N	6S	18S	30S	42S
Decrease in illum., %	100	100	100	100	100	100	100	84	27	4	0.6	0.2

In conclusion we should note that the following calculations show that the hydrodynamic transportation of dust and soot to the south is significant for the long-term dynamic of the climatic system following the large-scale pollution of the atmosphere of the northern hemisphere. Three months after the beginning of a nuclear conflict the upper layers of air above Antarctica would be clouded. Following the clouds of soot, the "nuclear winter" would spread over the mainlands of the southern hemisphere. The air temperature above land masses in latitudes up to 30° S would fall by 1–4 °C. The temperature in the tropics would be much lower than if calculated without considering the transportation of aerosols.

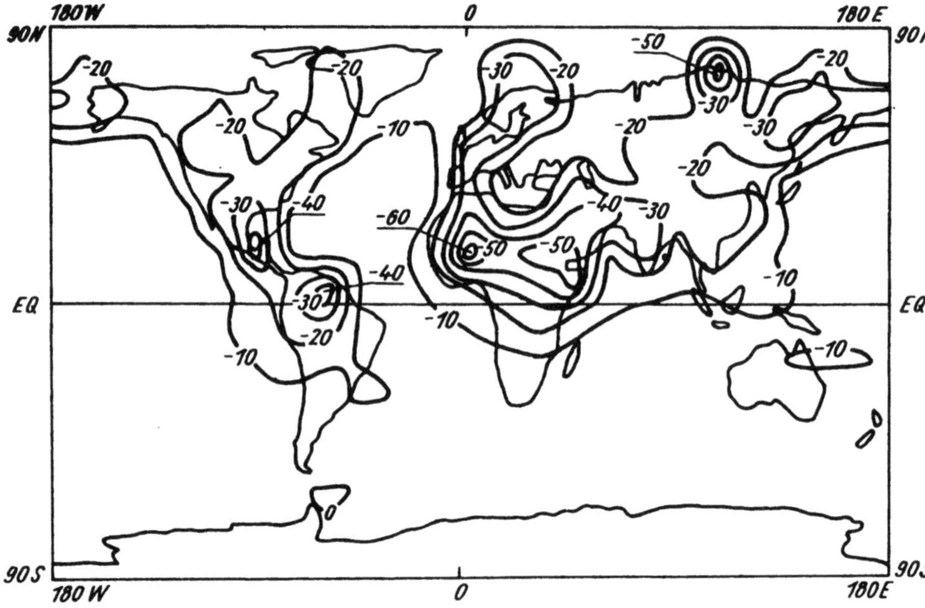

Fig. 8. Decline of air temperature at the surface of the earth from the scenario with the total yield of 10,000 Mt 40 days after the nuclear war

4.3. Mechanisms of low temperature impact on plants

Let us first consider the effect on plants of only one factor — low temperatures. To determine the effect of this factor we can divide all plants into those sensitive to cold and those sensitive to frost. Plants sensitive to cold suffer damage at temperatures above 0 °C; plants sensitive to frost suffer damage at temperatures below 0 °C.

Plants sensitive to cold include many species of tropical rain forest trees, various C_4 grasses, herbaceous dicotyledonous tropical plants and other plants, for which lethal temperatures range from 0–10 °C. The main cause of death in these plants is the change in cell membranes from a primarily liquid crystal state to a gel state as the result of the induration in lipids in the membrane [17]. The increase in the size of calluses in the membrane in the form of gels and the deformation of membrane protein leads to pathological changes in the transport of matter among cell compartments and in the basal metabolic rate of the cell, finally leading to the death of the cell. The temperatures required to bring about these phenomena depend on the content of unsaturated aliphatic acids and the chemical composition of phospholipids in the biomembrane [97].

The cause of death in plants sensitive to frost is the formation of ice inside the cell or in intercellular spaces. In the latter case, there is a great difference in the pressure of water vapor between the ice in the intercellular spaces and the liquid in the cell, leading to the dehydration of the protoplasm, its reduction in volume (up to 60 % of its normal value) and an increase in the concentration of substances dissolved in the cell. The death of the plant results from cell damage by dehydration and deformation caused by the formation of ice in intercellular spaces, denaturation of protein as a result of osmotic loads, and the increase in concentration of cellular solution to a toxic level.

Cold-resistance and frost-resistance are not constant properties of plants, but rather brought about according to genotype by the process of onthogenesis under the influence of external factors.

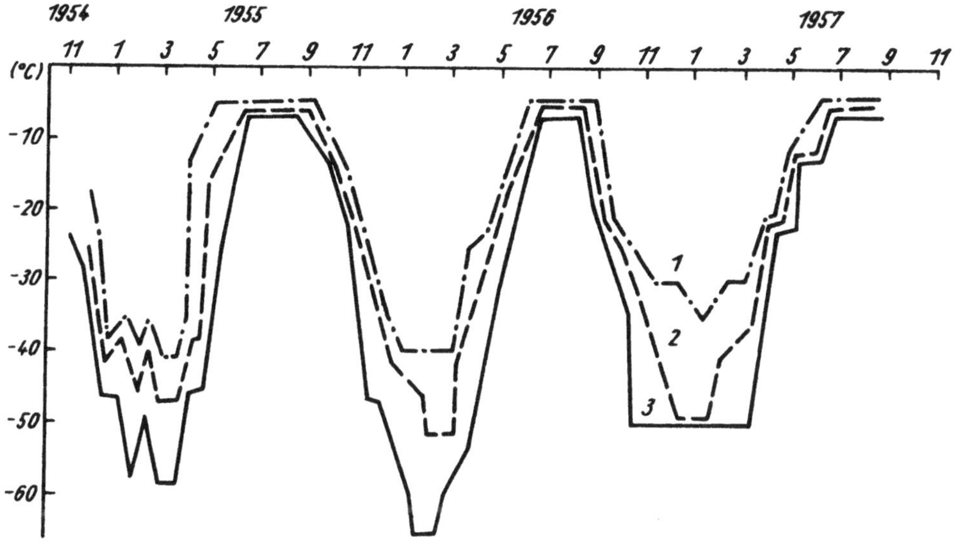

Fig. 9. Variation of the frost-resistance of trees during 3 years of observations
1 – apple-tree, 2 – spruce, 3 – birch [45]

Frost-resistance changes drastically in the course of a year. Figure 9 depicts the change in frost-resistance of three species of trees over three years of observation. We can see that frost-resistance is at a minimum during the summer and at a maximum during the winter.

According to I.I. Tumanov [72], plants must pass through three successive stages of preparation in order to become frost-resistant: entering a state of physiological dormancy, and passing through the first and second stages of tempering. High frost-resistance in a plant does not come about immediately, but rather increases by stages: first in a period of dormancy after the end of the vegetation season, then in a stage of tempering and, finally, in the second phase of tempering as a result of the slow and gradual increase in frost. Maximal frost-resistance is reached during the most severe season.

Each of these stages serves as preparation for the next. If any link in rather long (several months) preparation of a plant for the winter is omitted or is unsatisfactory, the plant will not be capable of acquiring the necessary frost-resistance.

For example, because of summer droughts, fruit plants often do not succeed in normally completing vegetation. For this reason they do not have time to prepare for winter, and die when hit by frosts that they would be able to withstand with no trouble after favorable summers. The Siberian firs can withstand frost down to –60 °C near the coldest point in Siberia, where they form extensive forests, yet they freeze on the banks of the Rhine in the warm climate of Central Europe, i.e. the external conditions necessary for the development of high frost-resistance are not always present.

Lethal temperatures for non-tempered plants that are sensitive to cold range from 0–10 °C, and for tempered plants from 0–5 °C. Lethal temperatures for non-tempered plants that are sensitive to frost range from (–1)–(–3) °C, and for tempered plants from (–3)–(–50) °C [17].

In the process of evolution plants adapted to the change in season and the ensuing lower air temperature, or to the onset of the dry season. In many perennials, during the process of transition to a state of dormancy, growth almost ceases, and the intensive accumulation of sugar begins.

Then the photosynthesis ceases completely or almost completely and, for deciduous plants, the leaves fall off. During the process of tempering in plant cells, defensive agents (cryoprotectors) accumulate, in the form of sugar, water-soluble proteins and organic acids. The unsaturation of lipids is also increased [34], and complex conformative changes of proteins occur. In trees a substantial (up to 40–50 %) decrease in the water content in the trunk is observed. All these changes enable plants to withstand lower temperatures.

4.4. The influence of decreased illumination on the transition to a state of dormancy

In order for plants to pass over into a state of dormancy, a sufficient level of illumination in the beginning stage of this process is essential. All processes connected with the transition to a state of dormancy require certain expenditures of energy. Assimilates produced by photosynthesis are the source of this energy. In the course of the vegetation period until the beginning of the transition state plants produce almost no mobile energy reserves, and assimilates formed by photosynthesis are used in the growth of biomass and the maintenance of life. The transition to a state of dormancy begins with a cessation of growth, and assimilates begin to accumulate – in the form of sugar – mobile energy sources.

Therefore, if illumination in the beginning of the transition state of dormancy is too low, plants will not receive enough energy for the necessary changes to take place. Under normal conditions illumination is somewhat lower toward the end of the vegetation period, but it is sufficient for the transition to a state of dormancy.

If illumination is significantly reduced (20–100 times in comparison with the light saturation level for photosynthesis), the amount of energy received is reduced to a point where it does not meet expenditures for respiration, and pure photosynthesis vanishes. This so-called compensatory intensity of illumination I_k is different for various species of plants, and decreases with air temperature.

In general, under normal conditions I_k is greater for photophilous and thermophilous plants (10–30 Wt/m^2) and less for shade resistant plants (1–5 Wt/m^2) [45]. For example, for the cotton plant I_k = 20 Wt/m^2 [54]. This is an illumination 25 times less than the light saturation level for photosynthesis. If illumination falls below a value of I_k, the plant may die.

It should be noted that plants of the same species in the same phytocoenosis may react differently to the impact of lower temperatures and illumination. Weakened trees, old trees and very young trees are less able to withstand these factors than others. Therefore, even if lowered temperatures and illumination do not reach the limits which would bring about the death of most plants, some portion of the plants will still die. The closer decreased temperatures or illumination come to these limits, the greater the percentage of trees that will die.

4.5. Estimations of death of vegetation and animals

We will consider two extreme cases: first, if a nuclear conflict were to begin in July; second, if the conflict were to begin in January. Absolute values of the air temperature on the underlying surface for various parts of the earth can be obtained by subtracting the calculated temperature reduction from the standard average summer values (if nuclear war were to begin during the summer). Values for a nuclear war beginning in winter can be determined analogously.

July is the warmest month in the northern hemisphere: According to the calculations, within 15 days after the spreading of pollution over the northern hemisphere, the air temperature at the land surface would fall below zero for almost all of the northern hemisphere. A zero isotherm would pass through the equator. On the ninth day after the spread of pollution, illumination at latitudes above 18° N would be less than 3.6×10^{-5} Wt/m^2. The given illumination is $3–80 \times 10^4$ times less than the compensatory illumination of plants, I_k, measured under normal temperatures. The taking up of energy by plants can be considered to have ceased. Would the plants have time to adapt to the low temperatures? It can be asserted that they would not.

For plants in the northern and central zones, the transition time to the dormancy state under normal conditions for the end of a vegetation season is more than two weeks. The main factor causing the transition to a state of dormancy is the shortening of day length. The reduction in air temperature has less influence. A minimum of 3–5 days passes from the initial effect of these factors to the transition to a regime of sugar-storing. If during this time, under the conditions of the incipient "nuclear winter", the mechanism for transition to a state of dormancy were triggered, plants would not have time to accumulate a sufficient quantity of assimilates (not more than 10 % of the necessary quantity would be accumulated) as a result of the rapid and severe reduction in illumination, and the transition to a state of dormancy would not occur. The subsequent negative effect of low temperatures in the course of more than three months would necessarily lead to the death of the plants.

Similarly, subtropical plants would not have time to make the transition to a state of dormancy and would die under conditions of low temperatures and the absence of light. Those plants in which the transition to a state of dormancy is linked with the onset of a dry period would die out. In tropical rain forests the illumination would be greater than the compensatory illumination (70 Wt/m^2 on the 40-th day, and 50 Wt/m^2 on the 99-th day after the beginning of a nuclear war), but since plants in these forests do not possess dormancy mechanisms, they would die from the effects of low temperatures.

In July it is winter in the southern hemisphere. The reduction in temperature for latitudes of 0–12° S would be 1–4 °C, and illumination (see Table 1) would be decreased by 30 %. Under such conditions not all plants could withstand an extended reduction in temperature or, more importantly, in illumination. In this case, weak, old and young trees would be primarily affected as described above. The zone in question is occupied mainly by tropical forests. Approximately 60 % of the plants in the upper layer of these forests are in a climax state; growth is absent and photosynthesis is equal to respiration. A reduction in illumination would lead to an energy deficit, for which the given plants would be incapable of compensating.

Shade-resistant plants would gain an advantage instead of the plants that had died. These plants would hinder the growth of young photophilous plants located underneath their canopy (15 % of all photophilous plants). Young photophilous plants would die as a result of being shaded by shade-resistant plants. By considering the relationship between photophilous and shade-resistant plants in tropical forests, we find that on the whole about 50 % of all plants would die.

In latidudes below 12° S, the reduction in temperature would not exceed 3 °C, and the reduction in illumination would not exceed 4 %. This would not cause significant damage to plants.

In all zones plant seeds would be preserved. Germination of seeds that fell to the earth would occur in the course of a few years. Therefore, after the end of the "nuclear winter", seeds that had matured and fallen to earth before it began would be able to germinate in the course of a few years. However, approximately 15–20 % of plant species would become extinct (according to expert estimates of biogeographers).

Thus, if nuclear war were to begin in July, all vegetation in the northern hemisphere would be killed, and vegetation in the southern hemisphere would partially decline to extinction (see Fig. 10).

The death of animals in the northern hemisphere under the given circumstances would be caused by lack of food and complications in the food search under the conditions of the "nuclear night". In tropical and sub-tropical regions, cold would be the most important factor. Many species of mammals and all birds would die, while reptiles could survive.

January is the coldest month in the northern hemisphere. Plants in the northern and central zones would be in a state of dormancy. Therefore, their ability to withstand a "nuclear winter" would be determined by the amount of frost.

The greatest decrease in temperature – up to 54 °C – would take place in latitudes of 12–36° N. The absolute temperature value in this zone would be (–6)–(–42) °C. The decrease in temperature in the far north in latitudes from 48–62° would be 11–38 °C. Here the absolute temperature value would be (–15)–(–72) °C.

In this case let us consider the effect of the "nuclear winter" individually for various types of vegetation. We will adhere to the biogeographical principle of analysis.

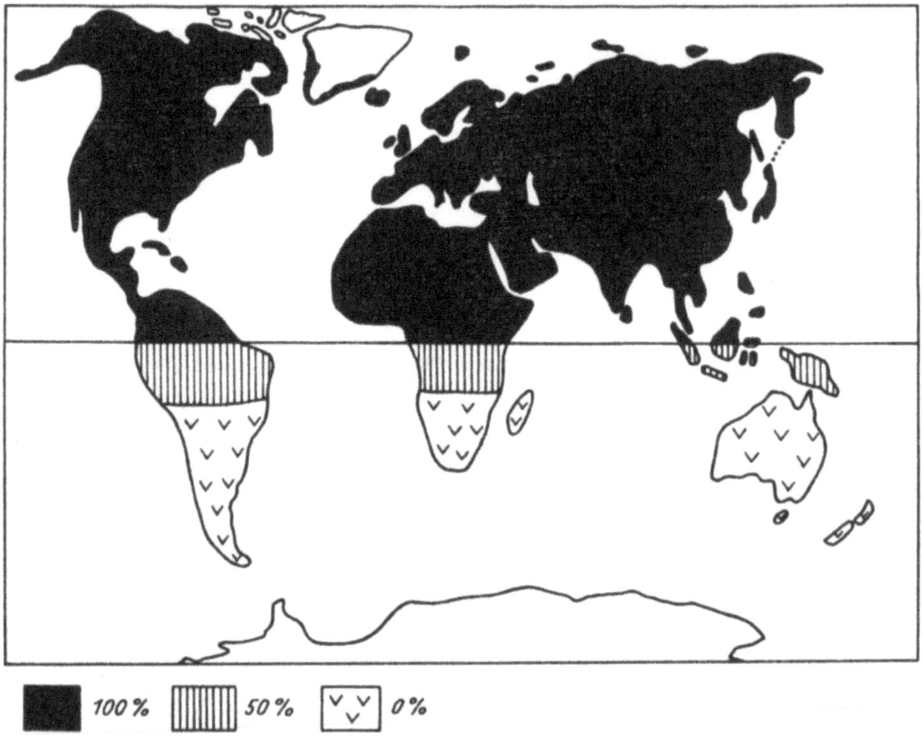

Fig. 10. Percentage of killed vegetation under the impact of the factors of the "nuclear winter" if the conflict begins in July

– 100 %; – 50 %; – 0 %

4.5.1. Tundra, forest tundra, taiga forests, wide-leaf forests

The distribution of individual tree species was compared with average absolute minimum temperatures in order to estimate the tolerance of the given plants to withstand frost [2]. An analysis of these data allowed us to estimate the minimum temperatures that can be tolerated by these trees during the winter (see Table 2).

Table 2. Maximum frost-resistance of trees (for 1 month of frost)

Trees	Beech	Oak	Birch	Spruce	Pine	Fir	Cedar	Larch
Frost-resistance in winter, °C	–25	–40	–55	–65	–65	–65	–55	–65

By comparing normal winter temperatures, temperatures during the "nuclear winter", and the proportions between species of plants, and by using data collected from observing the death of trees during anomalous winters, it was possible to estimate the percentage of tree death for each portion of land. This procedure was carried out as follows. The percentage of plant death during normal winters, corresponding to conditions averaged over many years, was compared with the

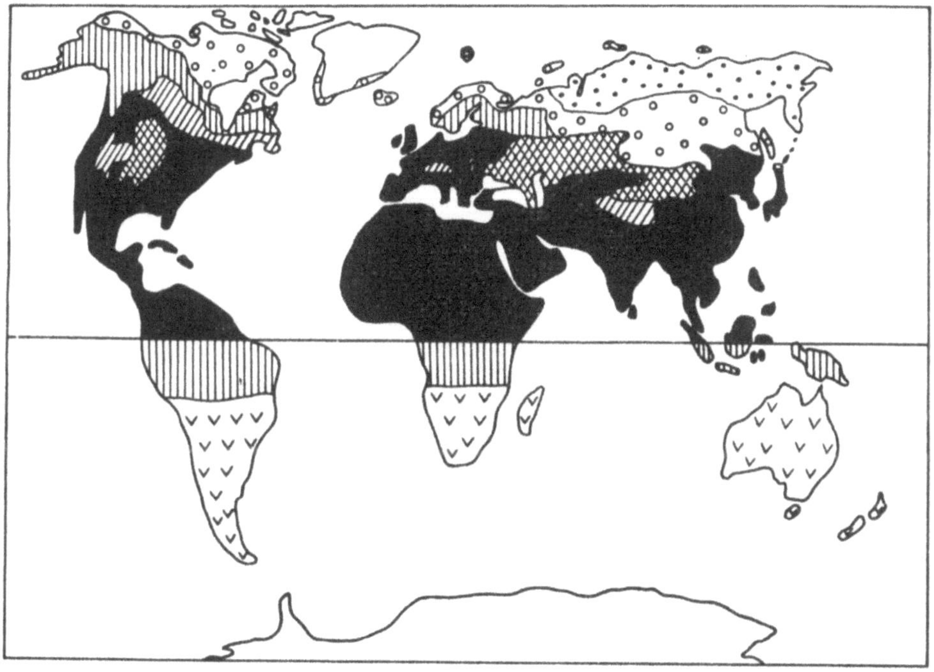

Fig. 11. Percentage of killed vegetation under the impact of the factors of the "nuclear winter" if the conflict begins in January

– 100 %; – 90 %; – 75 %;

– 50 %; – 25 %; – 10 %; – 0 %

percentage of plant death during anomalous winters with extended frost. Data on the percentage of plant death under the effects of frost during a "nuclear winter" were obtained by linear extrapolation of the above figures.

It was found that because of different temperature values during a normal winter and different temperature reductions during the "nuclear winter", the levels of death for the same plant species would be different in Europe, Siberia and North America. Data on plant death in these three regions are contained in Table 3 and Figure 11.

The corresponding territories occupied by the forest communities shown in Table 3 are indicated on the map of types of vegetation communities.

4.5.2. Grasslands

Cold temperatures in the grassland zone lead to the death of surface plants and to the almost complete freezing of plant root systems. Table 4 contains data on frost-resistance for various parts of grasses.

Frost resistance is (–11)–(–20) °C (see Table 4), while temperatures in grassland zones during the "nuclear winter" would be approximately (–23)–(–30) °C. Over the course of several months it is possible that a number of bulbous plants would survive. Approximately 90 % of the plants would die.

Table 3. Death of vegetation in several types of vegetative communities (in percent) (for nuclear war beginning in January)

Type of vegetation	Europe	Siberia	North America
Arctic desert, tundra	25	10	25
Forest tundra, northern taiga forest	25	10	50
Central taiga, southern taiga forest	50	25	75
Wide-leafed-coniferous, wide-leaved subtropical forests	100	100	100
Grasslands	90	90	90

Table 4. Maximum frost-resistance of the above-ground organs of grasses [72]

Plant part	Frost-resistance in winter
Winter-green leaves	
3–5 cm above leaf litter	from −11.5 to −14.5
5–10 cm above leaf litter	from −11.5 to −18.0
10–20 cm above leaf litter	from −13.0 to −20.0
Buds	
below leaf litter	from −7.0 to −11.5
directly above leaf litter	from −12.5 to −18.0
3–20 cm above litter layer	from −15.5 to −19.5

4.5.3. Mountain deserts, alpine and subalpine meadows

This type of mountain vegetation is adapted to withstand significantly low temperatures. Therefore, these plants would be partially able to tolerate the "nuclear winter". Approximately 75 % would die. In the alpine and subalpine meadows of Tibet, the decrease in temperature would be more than 50 °C, and nearly all vegetation would die.

4.5.4. Tropical and subtropical forests, savannah

The ability of these types of vegetation to withstand factors of the "nuclear winter" would be the same as for a nuclear conflict occuring in July. Therefore, plant extinction would be almost total.

4.5.5. Vegetation in the southern hemisphere

In January it is summer in the southern hemisphere. The ability of plants in the equatorial zone to withstand factors of the "nuclear winter" at this time would not differ significantly from the case of a nuclear war begun in July, since the difference in temperature between winter and summer in tropical zones is quite small. In the more southern part of the hemisphere, the change in temperature and illumination would be quite small and the effects of the "nuclear winter" would be insignificant.

4.5.6. Death of agroecosystems

In the given situation it is possible that winter crops will survive. All the other wintering agrocoenosis (fruit trees et al.) would die in a "nuclear winter", as is shown by the results of a comparison of frost-resistance and absolute temperature values. The frost-resistance of winter rye is (−30) °C, while frost-resistance for winter wheat, depending on the sort, is (−16)–(−26) °C. The temperatures in a winter crop zone during the "nuclear winter" would be (−22)–(−40) °C. Consequently, a small part (about 10 %) of winter crops may survive the frost. However, the probability of their survival after the end of the "nuclear winter" is practically zero, due to effects of other nuclear war factors.

If a nuclear war were to begin in January, the death of animals in the northern hemisphere would be brought about by severe cold and complications caused by low illumination in finding enough food to maintain energy requirements, which would increase under conditions of low temperatures. The death of mammals and birds under these conditions would be total. The death of animals in tropical zones would be approximately the same as for a nuclear war beginning in July.

4.5.7. Ocean

The ocean is the most conservative bloc of the biosphere. Because of its size, it has a damping effect on many local oscillations in climatic and biogeochemical factors. The "nuclear winter" would probably have the greatest influence on its ecosystems. However, during a "nuclear winter", the surface layer of the ocean would not be significantly cooled. Therefore, the main factor having an influence on ocean biota would be the decrease in illumination and complete cessation of photosynthesis. There would be a significant decrease in the number of phytoplankton, but they would not die out completely, since many species would become dormant and would thus survive the "nuclear winter" After the "nuclear winter", the number of phytoplankton would be restored in the course of a few years. It is possible that many species of fish would die out, mainly because of the lack of sufficient food. However, the complete collapse of the trophic pyramid would not take place, since bacterioplankton and dissolved organic substances would remain intact in the food chain.

4.6. Conclusion

The "nuclear winter" would have the strongest impact on land ecosystems. If a war were to begin during the summer, most vegetation of the northern hemisphere would freeze. Tropical vegetation would be destroyed in any case.

The resulting enormous areas of dead trees would serve as good fuel for secondary forest fires. The decay of this dead organic matter would lead to the emission of large quantities of carbonic dioxide into the atmosphere, which would seriously disturb the global cycle of carbon. The elimination of vegetation (especially in the tropics) would bring about active processes of soil erosion.

Practically all species of mammals and birds would die out.

The "nuclear winter" would be seriously detrimental to agroecosystems. All fruit trees, vineyards, etc., would be frozen. Almost all populations of farm animals would die out, since the infrastructure of animal husbandry would be destroyed.

The renewal of parts of the vegetation is possible (seeds would be preserved), but this process would be retarded by the impact of other aspects of nuclear war.

5. "ORDINARY" POLLUTION

5.1. Introduction

Currently much is said and written about the ever-increasing pollution of the environment. The emission of oxides of nitrogen and sulphur leads to the phenomenon of "acid rain", where rain and snow water in many countries consists of solutions of nitric and sulphuric acids with a pH = 3–5 (the pH of pure unpolluted water is 6.5–8.5). This results in the degradation of ecosystems (especially of coniferous forests). The degradation is a result of the change in the rate at which nutrients are carried out of the leaves of plants, and the acceleration of the leaching of calcium and other nutrients from the soil (soil fertility decreases).

If the oxides of nitrogen and sulphur remain in the atmosphere, the presence of hydrocarbons and the effect of UV-radiation causes them to form toxic substances such as ozone, formaldehyde, aldehydes, and peroxiacetylic nitrates (PAN) – components of photochemical smog. Plants (especially conifers) are most sensitive to SO_2, ozone and PAN.

However, while ozone is a highly toxic pollutant above the surface of the earth, the "ozone screen" in the upper layers of the atmosphere is essential for protecting the earth from excessive radiation. Oxides of nitrogen emitted into the stratosphere destroy this layer.

Quantitative information. Every year 5×10^7 tons (t) of oxides of nitrogen and 1.5×10^8 t of sulphur dioxide are emitted into the atmosphere as products of combustion.

Although natural emissions of oxides of nitrogen (7×10^8 t/yr) are an order higher than anthropogenic emissions, the impact of the latter on the biota has already become global. This fact demonstrates just how fragile and finely balanced the global biogeochemical cycle of nitrogen is.

Other rather serious toxic pollutants are the heavy metals (mercury, lead, cadmium, zinc, selenium, copper and arsenic*). Even in relatively low concentrations they inhibit the processes of growth and microbiological processes of the decay of dead organic matter, sharply reducing the rate of the natural cycles of carbon, nitrogen and phosphorus, and thus the productivity of the ecosystem. For example, a concentration of copper of 5×10^{-4} g per gram of soil decreases the phosphatic activity and rate of mineralization of nitrogen by half (in the surface layer of soil in coniferous forests). At present, heavy metal pollution has already a noticeable impact on many ecosystems.

Quantitative information. The yearly emissions of heavy metals (according to 1971 data) in thousands of tons are as follows: mercury – 2; lead – 75; copper – 60; zinc – 1100; arsenic – 1300; cadmium – 17 [74].

Finally, there is the pollution by petroleum, which mainly affects ocean shelf ecosystems. Petroleum kills off practically all life, especially in the upper and middle part of the litoral zone. The process of renewal in such a "dead zone" takes from five to ten years. Strangely enough, there exists natural petroleum pollution caused by rifts in the ocean floor. Man has now doubled the amount of this pollution through accidents involving tankers and off-shore oil wells (30 % of total anthropogenic pollution) and as a result of the disposal of petroleum products. There is one more risk factor from oil pollution, which unfortunately has received little attention. Oil pollution of the oceans in the central and tropical latitudes (of the northern hemisphere) are carried by current systems into the Arctic Ocean, where they are then circulated in a clockwise direction as a separate layer. As the upper layers of ice melt, this layer would rise until it reaches the surface of the ice. A sharp change of the albedo in the polar ice-caps would accelerate the melting-process, which could have a serious impact on the global climate.

Quantitative information. Every year 3×10^7 t of petroleum end up in the oceans of the world. Approximately 1 % of all petroleum being transported ends up in the ocean. One ton of petroleum forms a patch on the ocean surface with an area of 12 km^2 [39].

* Arsenic pollution will be included in the general concept of "pollution by heavy metals"

The oceans of the world are already seriously polluted by petroleum.

There is one more important source of anthropogenic pollution of the environment – poisonous chemicals (pesticides and herbicides). In contrast with ordinary chemical weapons (neuro-paralytic poisons), these substances (mainly organo-chlorides) are highly stable (a half-life of several years). Most chemical substances used in Vietnam were ordinary herbicides, but the concentrations in which they were applied (5–30 kg of active ingredient per hectare) were significantly greater than what is normally used in forestry and agriculture. The ecological consequences of such an application of herbicides were quite substantial (complete elimination of the plant cover over large territories and the exposure of the soil) [120].

Quantitative information. The volume of pesticides introduced into the soil in a year is 1.25×10^6 t. The average dose in Western Europe is 0.3–3 kg/hectare/year [39].

Accidents at poisonous chemical plants present a great potential danger. The main component of many of this type of substances is 2,3,7,8-tetrochloridibenzol-p-dioxin (TCDBD). The accident that occurred at a plant in Seveso (Italy) in July 1976, in which 2.5 kg of TCDBD where dispersed over several hundred hectares, leading to an ecological disaster, is indicative of the significance of this risk factor. A similar catastrophe in Bhopal (India) in December 1984 led to a regional disaster, in which thousands of people were killed and tens of thousands were disabled. A lethal dose of TCDBD for mammals is comparable to a dose of neuro-paralytic poisons, although it takes effect much more slowly; in sublethal doses it causes a large number of chronic toxicological illnesses, including congenital defects [120].

After a nuclear war, as a result of the destruction of industrial plants, fires in cities, forest fires, the destruction of oil works and oil reserves etc., the amount of "normal" pollutants emitted into the environment would be so great that it would have the most serious negative impact both on ecosystems and on man. In absolute terms this would be many times the normal yearly emissions, but since this would all occur during a period of several days, the intensity of these emissions would be increased hundreds and thousands of times. The detrimental effects of pollution would rise correspondingly – the "shock dose" effect. The principle of amplification would be manifested, and the picture after a nuclear war (in terms of the "ordinary" pollution factor) would recall a picture of the current "polluted" world, intensified many times over.

After these qualitative analyses, let us move on to quantitative estimations; we will attempt to estimate the level of "ordinary" pollution after a nuclear war and its impact on the environment.

5.2. Estimations of the emission of heavy metals and oxides of nitrogen and sulphur as a result of a nuclear war (without considering the burning of vegetation)

Pollution by heavy metals, as well as by some compounds of nitrogen and sulphur, which may be emitted into the atmosphere as a result of fires and the destruction of industrial plants after nuclear attacks, would play a significant role in the change in the nature of processes that take place in natural ecosystems. Among heavy metals the greatest danger is presented by the most toxic ones, having a relatively high degree of distribution, such as mercury, lead, cadmium, arsenic, selenium and zinc. Among compounds of nitrogen and sulphur, the most significant are oxides of nitrogen and sulphur: NO_x and SO_2. These compounds may be formed in chemical reactions of substances containing nitrogen and sulphur that are emitted from the surface of the earth as a result of fires. The same factors would lead to an excess concentration of heavy metals. They may be sublimated or even burned as a result of the increase in temperature, or they may be liberated as a result of the combustion of substances containing them. From this standpoint it is convenient to divide their sources into two basic groups:

1. *sources of the first type* – substances containing heavy metals obtained in processing polymetallic ore with the goal of extracting these metals;
2. *sources of the second type* – combustible substances containing the elements we are interested in.

Generally speaking, some substances may fall into both groups. Therefore we will agree to include in the second group only those substances which do not fit into the first group, i.e., substances in which the given elements or their compounds are contained in natural admixtures. All sources for compounds of nitrogen and sulphur will be considered to belong to the second group.

Now let us describe the release mechanisms of the elements concerned from sources of both types.

Estimations of the release of these elements from sources of the first type are made difficult by the extreme complexity of estimating the percentage of supplies that would be susceptible to emission into the atmosphere. To solve this problem it is necessary to be acquainted with the mechanisms of the accumulation and burying of heavy metals, which in turn may come to light only as a result of a detailed analysis of technological processes that involve these elements. We will maintain the point of view that the given metals are mined mainly for technical purposes, and that they will therefore be accumulated mainly in regions with a significant degree of urbanization. We will further hypothesize that the release of heavy metals into the environment as a result of a nuclear conflict would be caused first of all by the destruction of insulation (for elements such as mercury), and by the sublimation and combustion of these metals under conditions of sharply increased temperatures in areas bombed directly, and of fires caused by nuclear explosions. We will consider the amount of heavy metals emitted from sources of the first type only under the conditions of a global nuclear conflict to be equal to 10 % of the amount mined in the average year.

Sources of the second type include bitumens, benzenes, plastics and other petroleum products, as well as coal, peat and similar combustibles concentrated near the locations of nuclear explosions. Considering that these regions will be mainly densely populated cities, it is natural that we should estimate these sources for urban territories.

The main difficulty in estimating sources of the second type lies in determining the percentage of the given elements per unit of mass of the burning substance. This is connected with the nonuniformity of sources of the second type. However, the greatest contribution for the various elements would be made by materials containing them in the greatest quantity. Coal is one of such materials. Let us use coal as a standard for the estimation of the release of these elements by sources of the second type. We should note that a selective comparison shows that the content of these elements in various flammable substances is of the same order of magnitude. Thus, for example, the percentage of lead in the ash of coal coincides exactly with the percentage of lead in gasoline – 0.02 %. Note that ash mass on the average comprises 10 % of the total mass of coal.

Data relating to the content of toxic elements in coal are contained below [74] (see Table 1).

Table 1. Emission of heavy metals into the atmosphere as a result of coal combustion

1.	Element	Hg	Pb	Cd	As	Se	Zn	Cu
2.	Expansibility of steam at 2000 °C, atm.	200	2	100	150	120	60	0.04
3.	Content in emissions (g per 1 t of ash)	5	200	44	3540	–	3000	150
4.	Content in gases/ content in coal, %	90	50	85	86	–	80	4
5.	Emission from combustion in 1971 (thousands t)	1.9	76	16.7	1300	–	1100	40
6.	Extracted in 1971 (thousand t)	9	2800	13	40	1	3800	5400

As we can see from the table, all the given elements have a high expansibility of steam, which supports our hypothesis that they may be sublimated in zones of high temperatures. In any case, sources of the first type are the only sources of selenium, which is found in significant quantities neither in coal nor in petroleum.

In order to calculate the release of the elements concerned from sources of the second type, we should first estimate the quantity of conventional fuel burned as a result of fires, for the parameters of which we use the parameters of coal, as noted above. Let us first estimate the present per capita supply in areas of possible nuclear attacks.

According to [95, 105], in cities 1–20 g of conventional fuel are found in 1 cm^2 of area, which corresponds to 1–20 t/ind. in cities with the population density of Moscow (8 X 10^6 ind. in 800 km^2). If we consider the fact that in the regions that would suffer the main nuclear attacks, urban population makes up 50–90 %, we can consider this to be the given quantity of flammable substances for each inhabitant. For our calculations we will hypothesize that 10 t of the given substances would be burned for each person, with the conflict covering a territory with a population of 1 billion people. Thus we estimate the total amount of burned substances at 10^{10} t. We will consider the concentration of heavy metals (and of nitrogen and sulphur) in these substances to be the same as in coal. Sources of selenium are of the first type only, as mentioned above. Data obtained on emissions are summarized in Table 2, which also contains estimations of emissions of hydrocarbons, nitrogen and sulphur.

This table was created in the following way. Let N_a be equal to the content of a given element in coal, calculated per unit of ash mass (these values are shown in line 3 of Table 1). Then, in accordance with the hypothesis on the correlation between the mass of ash and the mass of coal, the amount found in coal would be $N_c = 0.1\ N_a$. These values (rounded off to the first significant digit), supplemented by data concerning hydrocarbons, nitrogen and sulphur (which were calculated according to their percentages in coal [25]), are contained in line 2. Line 3 gives estimations of emissions for the burning of 10^{10} t of conventional fuel, i.e. for "fires of civilization". Thus, in this line we find values of $10^{10}\ N_c$. Line 4 contains the data from line 5 of Table 1 reduced by a factor of 10, which corresponds to our hypothesis on the combustion and sublimation of heavy metals from sources of the first type (in quantities equal to 10 % of yearly production). If we compare the values contained in lines 3 and 4 of Table 2, we can see that the amounts contributed by sources of the first and second types are approximately equal for lead and mercury; for cadmium, arsenic and zinc, pollution is mainly contributed to sources of the second type.

From the last line of Table 2 we can see that total emissions as a result of "fires of civilization" are equal to approximately three times the yearly norm of "ordinary" pollution for mercury, cadmium, arsenic and zinc, seven times the yearly norm for lead, and 12 times the yearly norm for copper.

Table 2. Estimations of the emissions of elements in "fires of civilization" (basing on data from Table 1 and the accepted hypotheses)

1	Element	Hg	Pb	Cd	As	Se	Zn	Cu	C	N	S
2	Content in emissions (g per 1 t of conventional fuel)	0.5	20	4	400	–	300	15.6	$9 \cdot 10^5$	$7 \cdot 10^4$	$3 \cdot 10^4$
3	Emissions from sources of the 2-nd type (kilotons)	5	200	40	4000	–	3000	156	$9 \cdot 10^6$	$7 \cdot 10^5$	$3 \cdot 10^5$
4	Emissions from sources of the 1-st type (kilotons)	4	300	1.3	4	0.1	400	540	–	–	–
5	Total emissions (kilotons)	6	500	40	4000	0.1	3000	700	$9 \cdot 10^6$	$7 \cdot 10^5$	$3 \cdot 10^5$
6	Annual emissions under normal conditions according to data for 1971 (kilotons)	2	75	17	1300	–	1100	60	$3 \cdot 10^6$	$2 \cdot 10^5$	10^5
7	Relative emissions (with respect to the yearly norm)	3	7	3	3	–	3	12	3	3	3

Undoubtedly the distribution of pollution over various regions is of interest. We assume that emissions of heavy metals as well as of oxides of nitrogen and sulphur are distributed in proportion with population density. We also assume that these emissions, rather settle to the surface of the earth than disperse over the entire world, thus making the intensity of the corresponding pollution proportional to population density. Of course these are rather rough hypotheses, but the orders of magnitude thus estimated are sufficiently exact. These calculations are contained in Table 3. The second line of this table contains the intensity of pollution by metals, as well as by nitrogen and sulphur (in the form of oxides), for territories where the average population density is equal to 1 ind./km². The total mass of emissions (see line 5 of Table 2) was calculated for the population in regions affected by the war (about 10^9 ind.), thus giving us absolute values for the amount of pollutants per person currently living in the region. The figures in the remaining lines were obtained by multiplying the figures in line 2 by the corresponding values of population density in the given regions.

Table 3. The post-war density of pollution for same regions (mass unit/km²)

1	Element		Hg	Pb	Cd	As	Se	Zn	Cu	N	S
2	Pollution of territories with population density 1 ind./km²		6 g	500 g	40 g	4000 g	0.1 g	4000 g	700 g	$7 \cdot 10^5$g	$3 \cdot 10^5$g
3	Region	Population density ind/km²									
4	Western Europe Japan	~200	1 kg	0.1 t	10 kg	0.7 t	20 g	0.7 t	150 kg	150 t	60 t
5	Eastern Europe N.Americ (east)	~ 20	0.1 kg	10 kg	1 kg	70 kg	2 g	70 kg	15 kg	15 t	6 t
6	N.America (west) Canada, Ural region	~ 2	10 g	1 kg	0.1 kg	7 kg	0.2 g	7 kg	1.5 kg	1.5 t	600 t

5.3. The effect of pollution by heavy metals on land ecosystems

Unfortunately, quantitative information on the effects of heavy metals on land ecosystems is so heterogeneous that we are forced to make only qualitative estimations. First a few general facts.

It is well-known that living matter on all levels of organization is not adapted for high concentrations of heavy metals in the environment. For example, concentrations on the order of 10^{-5}–10^{-6} g/h are sublethal for a number of organisms [106]. It has been determined experimentally that under concentrations of $PbCl_2$ of 1×10^{-6} g/h the intensity of photosynthesis in algae is reduced by 25 %, while for an extended impact (2–3 days) a concentration of only 1×10^{-7} g/h reduces photosynthesis by 25–50 %. High concentrations of lead, cadmium and nickel in the soil (10^{-5}–10^{-6} g/h) suppress photosynthesis and destroy the mechanism of somatic regulation in corn and sunflowers. An increase in the concentration of cadmium in the soil to 3×10^{-5} g/h leads to damages in plant tissues (op. cit. [74]).

Let us compare pollution by heavy metals after a nuclear war with the existing pollution of the soil (average analytical data and data on heavily polluted industrial areas which already clearly show the toxic effect of pollution on ecosystems). For our calculations we will take a layer of soil with a thickness of 0.8 cm and a specific gravity of 1.2 g/cm³ (upper-turf soils). This hypothesis is fully justifiable, since, first of all, decaying litter and the upper humus layer of the soil are subject to more intense accumulation of heavy metals (in comparison with other components of the ecosystem); secondly, the rate of vertical migration of salts of heavy metals is very low (1–2 cm per year). All emissions of heavy metals are assumed to settle to the surface of the earth at the point of emission in proportion with population density. The results are contained in Table 4. Data on current concentrations are taken from [74]; information from 5.2. is used in calculating potential concentrations.

From this table we can see that for cadmium, for example, concentrations toxic for plants have

Table 4. Soil concentrations of metals after nuclear fires compared to the present concentrations (mkg/g); 1 mkg = 10^{-6} g

Metal	Hg	Pb	Cd	As	Zn
Soils (mean analytical data)	0.1	10	0.5	5	50
Regions affected by industrial pollution (Western Europe, Western USA)		42	1–10	7–650	5–2500
Regions in zones of industrial pollution (Japan, Western Europe)		160–1500	100		
Emissions in regions with population density 1 ind./km^2	$6 \cdot 10^{-4}$	$5 \cdot 10^{-2}$	$4 \cdot 10^{-3}$	$4 \cdot 10^{-1}$	$4 \cdot 10^{-1}$
The same for population density 200 ind./km^2	0.1	10	0.8	80	80

already been reached in industrial regions. "Nuclear" pollution will only raise the total contamination of the soil slightly (this is true for almost all metals, with the exception of arsenic). Arsenic contamination would be quite substantial. At first glance it would seem that "nuclear" pollution by heavy metals would not add significantly to the total contamination of the soil, and its negative effect on plants and soil flora and fauna would be slight. However, it should be remembered that the pollution picture is highly uneven, with pollutants falling first on plant leaves, i.e. in a zone of direct contact with living matter (when their toxic effect would be much higher), and only later being washed onto the surface of the earth.

There is one additional aspect of the problem: while we know quite a bit about the chronic impact of pollution of this type, in the case of "nuclear" pollution we would have an instance of "shock" pollution and an acute impact. We know very little about the consequences of such an impact. Judging by analogy with the impact of radiation, we can say that negative effects would appear at smaller doses from acute influences than from chronic ones.

Obviously the major effect of "nuclear" pollution by heavy metals would occur not on the level of land ecosystems, but on the level of freshwater ecosystems (see Chapter 8).

Finally, a few words about synergism. Synergism of "radiation contamination + heavy metal pollution" would be the most distinctive. Heavy metals (even in doses much lower than lethal doses) block processes of fermentation, thus sharply decreasing the rate of post-radiation renewal and, as a result, increasing the effect of radiation on living systems.

5.4. "Acid" rainfall

As a result of fires in cities (see § 1. of this chapter) and forest fires (see Section 3.4.), approximately 10^9 t of nitrogen and 3×10^8 t of sulphur would be emitted into the atmosphere in the form of NO_x and SO_2. "Fires of civilization" would make the greatest contribution to these values; it is therefore possible to determine with sufficient accuracy that the distribution of emission intensity would be proportional to the population density in the corresponding regions.

The average period of time that oxides of nitrogen remain in the atmosphere is 5 days for nitrogen, while oxides of sulphur remain there for 3 days [39]. Let us take the mean figure of 4 days. Let us assume that in the course of this period these oxides fall to the surface of the earth along with rainfall. This rainfall would consist of solutions of sulphuric and nitric acids.

If we know the total amount of oxides of nitrogen and sulphur, we can calculate the concentration of HNO_3 and H_2SO_4 in rainfall as a factor of population density in the regions concerned (without taking into account dispersion, taking into account dispersion of 50 % of the emissions over the entire northern hemisphere, and taking into account complete dispersion). For our calculations we decided to use gram-molecular coefficients of dissociation of nitric and sulphuric acids at 18 °C, equal to 82 % and 51 %, respectively (rounded off to 100 % and 50 %). Concentrations were calculated in relation to an average 4-day rainfall of 0.33 cm (for a yearly norm of 30 cm

Table 5. Densities of pollution by nitric and sulphuric acids, their concentrations in "acid rain" and pH of the rainfall

1	Type of estimate	Density of pollution (kg-moles/km^2)		Concentration (g-moles/l)		pH		total pH
		HNO_3	H_2SO_4	HNO_3	H_2SO_4	HNO_3	H_2SO_4	
2	In conditions of uniform distribution	200	40	$6 \cdot 10^{-2}$	$1.2 \cdot 10^{-2}$	4.2	5	4.1
3	With no dispersion (in regions with population density 1 ind./km^2)	50	10	$1.5 \cdot 10^{-2}$	$3 \cdot 10^{-3}$	4.8	5.5	4.7
4	With 50 % dispersion (in regions with population density 1 ind./km^2)	125	25	$3.8 \cdot 10^{-3}$	$7.5 \cdot 10^{-3}$	4.4	5.1	4.3
5	The same as in line 3 (in regions with population density 200 ind./km^2)	$1 \cdot 10^4$	$2 \cdot 10^3$	3	0.6	2.5	3.2	2.4
6	The same as in line 4 (in regions with population density 200 ind./km^2)	$5 \cdot 10^3$	10^3	1.5	0.3	2.8	3.5	2.7

Total emissions: HNO_3 – $3 \cdot 10^9$ t, $5 \cdot 10^{10}$ kg·moles; H_2SO_4 – $1 \cdot 10^6$ t, $1 \cdot 10^{10}$ kg·moles

in Europe and North America), which comes to 3.3×10^3 m^3/km^2. It is assumed that all emissions of nitrogen and sulphur would be formed into these acids. The results of these calculations are contained in Table 5.

From the table we can see that the acidity of rain would be significantly increased (to a pH of 4.1–4.7) even in territories that are relatively weakly affected by the war (territories with a low population density).

Rains in territories with high population density would be most "acid" (pH = 2.4–2.7). Rainfall with this pH would have a direct toxic effect on the leaves of plants, ending with their complete death.

The question of the acidity of the soil is much more complex, and the degree of its acidification depends largely on the type of soil and litter species. It is therefore difficult to estimate this effect quantitatively. The average acidity of soil in zones of "acid" rain may be increased 2–10 times. What are the possible consequences?

It is known [45] that a pH of less than 3 causes severe harm to the protoplasm in root cells of most leaf stalk plants. In addition, the ions of many metals (aluminium, copper, zinc) are more easily mobilized in an acid environment, which has a toxic effect on the roots. In an acid environment living organisms accumulate a quantity of heavy metals an order of magnitude higher than in a neutral environment. This is a typical example of synergism of the type "pollution by heavy metals + acid rain". In acid soils, microbic decay of organic matter is disturbed and NH_4^+ is accumulated instead of NO_3^-, which leads to the disturbance of the normal biogeochemical cycle of nitrogen. Leaching from the soil of calcium and other nutrients is increased. This all leads to a decrease in the fertility of the soil.

Of course under conditions of highly industrialized agriculture these effects may be balanced by the use of fertilizers and deoxidizing additives, but if we take into account the fact that these industries will be destroyed as a result of a nuclear war, then large-scale "acid rainfall" could lead to a significant decrease in the fertility of the soil, making it unfavourable for raising agricultural crops. On the other hand, even a relatively small decrease in the pH of the soil (to 4–5) brings about conditions favourable for the development of ruderal species (various weeds).

Acid rain has the greatest influence on the biom of coniferous forests. If we proceed by analogy, we can say that as a result of the effect of this factor after a nuclear war we would be faced with a situation similar to that existing at present in Northern Europe, the northeastern USA and southeastern Canada, but intensified many times over.

Finally, a few words about freshwater ecosystems. A significant portion of the acid would end up in fresh water reservoirs as a result of surface drainage. The effects of the acidification of rivers and lakes is well-known (death of many hydrobionts, cessation of the development of fish roe, etc.). This is yet another negative effect of large-scale "acid rainfall" on the environment, an effect which would affect vast territories.

5.5. Emissions of carbon

Forest fires would be one of the main sources of emissions of carbon into the atmosphere (10^{10} t) along with "fires of civilization" (9×10^9 t). A total of 2×10^{10} t of carbon would be emitted into the atmosphere – almost 6 times the yearly norm of emissions of an anthropogenic origin. This value is comparable with the production of carbon of a natural origin $-6-7 \times 10^{10}$ t; consequently, it can be expected that "nuclear emissions" would have a significant influence on the basic biogeochemical cycle of the biosphere – the carbon cycle. On the other hand, because of the "greenhouse effect" these emissions would also have an influence on the working of the "climatic machine" of the earth over a rather extended period of time after the "nuclear winter" (see Section 7.7.).

But here we will be interested in a different aspect of the problem, namely the toxicological aspect. As a result of fires, mainly carbon dioxide, CO_2, would be formed. But in the lower layer of the atmosphere, where an oxygen deficit would occur as a result of large-scale fires, combustion would not be complete and, as a result, carbon monoxide, CO, would be formed. In addition in the presence of a large amount of water vapor, the reaction $CO_2 + H_2O = 2CO + H_2$ would play a great role. According to various estimations, approximately 10–25 % of the carbon would be used in the formation of carbon monoxide. We will use the mean estimation of 15 % (see Section 3.5.). As a result, 0.65×10^{10} t of CO would be formed in the atmosphere. If we assume that the carbon monoxide would be concentrated in the lowest kilometer of the atmosphere, and if we accept the hypothesis of the absence of atmospheric diffusion (and the distribution in proportion to population density), then the concentrations of carbon monoxide in territories of different population densities would be as follows:

in territories with a density of 1 ind./km^2 – 5×10^{-4} %,
in territories with a density of 20 ind./km^2 – 0.01 %,
in territories with a density of 200 ind./km^2 – 0.1 %.

CO is not very water-soluble, and the residence time in the atmosphere in high concentrations would be many months and years (residence time on the surface of the earth – 0.1–3 years [39]). It should be noted that when the atmosphere contains 0.02 % CO, a person loses consciousness after a few hours. A dose of 0.075 % is lethal [57]. If we compare these values with the above values, we can see that in densely populated territories the concentrations of CO would exceed the lethal dose. Even in territories of medium population density, the concentrations of CO would reach a level close to a dangerous level, which would cause loss of consciousness.

Of course these are maximal estimations, since the processes of atmospheric mixing would lead to a decrease in these concentrations. However this would not occur immediately, but rather in the course of a few days, and this time period would be sufficient for the toxic effects of CO to appear.

In conclusion it should be noted that such concentrations of CO have a significant inhibitory effect on photosynthesis.

5.6. Oil pollution of shelf zones after a nuclear war

Oil pollution of shelf zones would be one of the significant factors of nuclear war. An enormous amount of oil would be partially burned and partially discharged into coastal zones after attacks on seaports, through which oil transport from oil-producing countries takes place, and on off-shore oil rigs and oil pipelines, and after the destruction of oil and gasoline storage tanks. The consequences of the accident involving the tanker "Tory Canion", in which several hundred kilometers of the Atlantic coast of France were poisoned by oil, are well-known.

It is rather difficult to estimate the correlation between burned and spilled oil after the destruction of oil storage tanks and the like, but the likely hypothesis is that the relationship would be 2 : 1, i.e., 1/3 of the total oil supply in the coastal zone would be discharged into the ocean. Reports by the mass media on the consequences of the bombing of oil storage tanks in the region of the Persian Gulf during the Iran—Iraq conflict support this hypothesis.

The usual estimation of the characteristic residence time of oil in a port is 3–4 days. Thus an amount of oil equal to the daily norm of transported oil (lower estimate) will be spilled in the coastal zone as a result of the destruction of oil storage tanks of the port, accidents involving tankers in the port, and the destruction of pipeline terminals. If we take into account the fact that yearly oil imports in developed free-market countries is 10^9 t, then 3×10^6 t of oil would be discharged as a result of the destruction of the oil transport system alone (Fig. 12).

On the other hand, the amount of oil products that ends up in the ocean every year is 2 to 10 mill. tons. Thus "nuclear" oil pollution would be on the order of annual "ordinary" pollution and this factor would not seem to be very significant. But this is not so, for the following reasons.

Fig. 12. Main ways of oil transport. Thickness of arrows presents the fraction of the total oil transport
\\\\\\\\ -regions of off-shore oil rigs

First of all, annual "ordinary" pollution is "distributed" more or less uniformly over most of the world's oceans, while "nuclear" pollution would be localized along the Atlantic coasts of Europe and North America, the Mediterranean Sea and Gulf of Mexico, along the coasts of Japan and California, and so on.

Secondly, there is a substantial difference in time scales, i.e., a difference in the *intensity* of pollution. "Nuclear" pollution occurs over the course of days (1 t of oil spilled over 1 ha of area in the course of a day spreads over a water surface with an area of 12 km^2, forming a film with a thickness of 1.02×10^{-3} mm). Therefore, its intensity would be 300 times greater than the intensity of "ordinary" pollution. If we take into account the extreme non-uniformity of "nuclear" pollution, we must increase this value many times over.

It should be noted that the mechanisms of self-cleaning have time to take effect for "ordinary" pollution of low intensity. For example, oil gradually evaporates as a result of the interaction between water and the atmosphere: 15 % in the first day, and 20 % for the next 10 days [39]. Strong winds, which form foam on top of the waves in the open sea, allow particles of oil to be carried into the atmosphere along with water droplets. Under the influence of these factors the initially continuous film of oil is broken up, floculas are formed, these floculas are coated with microorganisms and the further destruction of the oil takes place. Photochemical oxidation of oil also plays a role. However, all these mechanisms take effect in the course of months and years, while the specific time period for "nuclear" pollution is one day.

The ecological consequences of catastrophic oil pollution of shelf zones is known too well for us to describe them in detail. Therefore we will present some estimations of the scale of oceanic oil pollution after a nuclear war.

It should be noted that while the destruction of off-shore oil industry is not a significant factor for Western Europe in comparison with the destruction of oil transport (this factor would bring about a significant increase in pollution only in the North Sea), the significance of this factor is increased for the USSR, which does not import oil, and for the USA, which imports a relatively small amount. Furthermore, in these countries there exists a large number of off-shore oil rigs (Gulf of Mexico, California, Alaska, the northern part of the Atlantic Coast in the USA; the Caspian Sea, Black Sea and southern part of the Pacific Coast in the USSR). For quantitative estimations we will use the same hypothesis as for the destruction of oil transport: an amount of oil equivalent to the daily norm produced would be discharged as a result of accidents involving oil rigs (Fig. 12).

Table 6 contains the results of calculations of possible oil pollution on the basis of the above hypotheses. It shows, on the one hand, the possible sources of pollution and their values (in terms of the possible polluted area); on the other hand, it shows the seas and regions of the aquatorias of the world's oceans adjacent to these sources (with indications of the area of sea aquatorias). In our calculations we have used the following information sources [52, 116].

From Table 6 we can see that almost all seas (and of course coastal regions) in the regions which would suffer from nuclear attacks would be covered by an oil film. The most serious situation would arise in European seas, in the Mediterranean Sea, on the Atlantic coast of the USA, and in the coastal waters of Japan.

Note that in the calculations contained in Table 6 we did not take into account the additional potential sources of pollution.

First, this includes regions of off-shore rigs not considered (for example, the Atlantic coast of the USA, Pacific coast of the USSR, etc.).

Secondly, the destruction of supertankers on major shipping lanes of oil transport. This can lead not only to an increase (quite significant) of oil pollution in the regions shown in Table 6, but also to the pollution of other regions of the oceans of the word which are not directly adjacent to the regions of nuclear conflict.

Final conclusion: a significant part of the shelf zones of the oceans would be polluted by oil. It should be noted that at present the shelf zone ecosystems are an important source of protein. For example, 90 % of the total fish catch occurs in shelf zones [39]. Pollution of these zones can lead to the destruction of this source. In the best instance the ecosystems of shelf zones would be renewed after 5–10 years; in the worst instance these ecosystems would be completely destroyed.

There is one further purely ecological aspect of this problem. The accumulation of toxic sub-

Table 6. Oil pollution of seas in regions of the nuclear conflict

Region, source of pollution	Annual import or annual production (mln. tons)	Possible polluted area (mln./km²)	Adjacent seas and aquatorias of the world	Area mln. km²
USSR:				
The Caucasus off-shore rigs	20	0.6	Caspian	0.3
North America:				
import	320	10	Atlantic coast	–
off-shore production in USA:				
California	2	0.07	Pacific coast	–
Louisiana, Texas	33.5	1.1	Gulf of Mexico	1.6
Western Europe:				
import	650	21	Mediterranean	2.5
			Bay of Biscay	0.4
			Baltic	0.4
			North	0.5
Great Britain:				
off-shore production	113	3.7	North	0.5
Norway:				
off-shore production	25	0.8	Norwegian	–
Japan:				
import	250	8.2	Sea of Japan	
			Pacific coast	1
Australia:				
import	17	0.6		

stances in sediments occurs as a result of physico-chemical processes of sedimentation and biological self-cleaning. These processes are most intensive in coastal waters, where the greatest species diversity, productivity and stability of ecosystems is observed. While rivers play the role of "collectors" of pollution, coastal zones are the "water treatment plants" of civilization. Because of the action of shelf ecosystems, the purity of the oceans is relatively well preserved. As a result of nuclear war a significant portion of these "water treatment plants" would be destroyed, leading to intensified pollution of the oceans.

5.7. Conclusion

As a result of a nuclear war, industries, oil rigs, oil storage tanks, etc. would be destroyed, and large-scale "fires of civilization" and forest fires would appear. Consequently, huge quantities of "ordinary" pollutants would be emitted into the environment. If we proceed by analogy, the current picture of the impact of "ordinary" industrial pollution is intensified many times over, but this would be a "shock" or acute impact.

Approximately 12 times the annual norm of copper, 7 times the annual norm of lead, and 3 times the annual norm of mercury, arsenic, cadmium and zinc would be emitted into the environment.

The total effect of "fires of civilization" and forest fires would lead to the emission into the atmosphere of 6–10 times the annual norm of pollutants such as oxides of nitrogen and sulphur, which cause a significant increase (up to a pH of 2.4–4.7) in the acidity of precipitation and, as a result, the acidification of the soil and freshwater bodies.

A significant amount of carbon monoxide and other pyrotoxins would be produced by fires.

Large-scale pollution of shelf zones would arise as a result of the destruction of oil rigs, oil storage tanks, pipelines, tankers, etc.

The above allows us to conclude that "ordinary" pollution after a nuclear war would have a significant impact on land and shelf zone ecosystems, causing their degradation, which, in turn, would lead to a substantial disturbance of global biospheric processes.

This factor would also have a serious impact on the human population.

6. ULTRA-VIOLET (UV)-RADIATION

6.1. Estimation of changes in UV-radiation

Immediately after a nuclear conflict an enormous quantity of oxides of nitrogen would be emitted into the atmosphere. If these oxides were released into the lower layers of the stratosphere, there would be a substantial impact on the "ozone screen" of our planet, which protects its biota from UV-radiation. A change in the thickness of the ozone layer would lead to an increase in the intensity of UV-radiation on the surface of the earth. It should be noted that this effect would be noticeable only after the end of the "nuclear winter", when the transparency of the atmosphere would be restored to its previous state.

Let us estimate the change in intensity of UV-radiation reaching the surface of the earth as a result of a decrease in the ozone layer. Ozone in the atmosphere combines with molecules of NO_x. Mainly a reaction with nitrogen dioxide (NO_2) takes place:

$$2NO_2 + O_3 = O_2 + N_2O_5 + 60 \text{ kcal}.$$

The amount of ozone in the atmosphere is commonly measured by the height h of a column of ozone under normal conditions of temperature and pressure on the surface of the earth. This value is $h = 0.3$ cm [19] (4×10^{37} molecules of ozone) on the average. The recorded fluctuations in the thickness of the ozone layer are $0.2 < h < 0.6$ cm [19]. Most ozone is located in the stratosphere.

According to the estimations in [28], approximately 10^{32} molecules of NO_x would be formed after a nuclear explosion with a yield of 1 Mt. According to the AMBIO scenario, the total yield of nuclear explosions would be 5742 Mt. Thus if the entire amount of NO_x formed by the explosions were to react with molecules of ozone, 2871×10^{32} molecules of ozone would be combined. Let us assume that all oxides of nitrogen would be combined with molecules of ozone. The amount of ozone in the atmosphere would thus be reduced to $397{,}129 \times 10^{32}$ molecules. This corresponds to a reduction in the height of the ozone column of 0.03 cm. The given value is an order of magnitude lower than the natural variability in the thickness of the ozone layer described above. The corresponding increase in the intensity of UV-radiation in a band of 200–300 nm would be 6 % of the average level. Consequently, nuclear explosions alone would not lead to a significant increase in UV-radiation. This is provided by a minimum estimation.

In order to obtain a maximum estimation, we assume that all nitrogen formed as a result of "nuclear" fires would be emitted into the stratosphere in the form of various oxides. According to the estimations in Chapter 5, approximately 7×10^8 t of nitrogen, i.e., $\approx 3 \times 10^{37}$ nitrogen molecules, would be emitted into the atmosphere. We assume that all oxides would combine with ozone. We also assume that the effect of other factors that might change the amount of ozone are insignificant compared to the impact of nitrogen oxides. Thus the thickness of the ozone layer would be decreased to $h = 0.1875$ cm. This is the maximum estimation.

Rays falling perpendicular to the surface of the earth would experience minimal attenuation. The following law of attenuation applies for this ray [19]:

$$S_\lambda = S_{\lambda 0} 10^{-k_\lambda h}, \tag{1}$$

where S_λ is the flux of radiation at the surface of the earth for a wave length λ (measured in $Wt \cdot m^{-2} \cdot nm^{-1}$), $S_{\lambda 0}$ is the flux of radiation for a wave length λ at the upper boundary of the stratosphere, K_λ is the attenuation coefficient (cm^{-1}), and h is the thickness of the ozone layer (cm).

The dependence of K_λ on λ is depicted in Figure 13 [64]. We can approximate this dependence by the following formula:

$$K_\lambda = \begin{cases} 2.4\lambda - 470, & \lambda > 250 \text{ nm}, \\ -2.4\lambda + 730, & \lambda < 250 \text{ nm}. \end{cases} \tag{2}$$

The dependence of $S_{\lambda 0}$ on λ is depicted in Figure 14 [64]. We can approximate it as a linear function:

$$S_{\lambda 0} = 0.0075\lambda - 1.75. \tag{3}$$

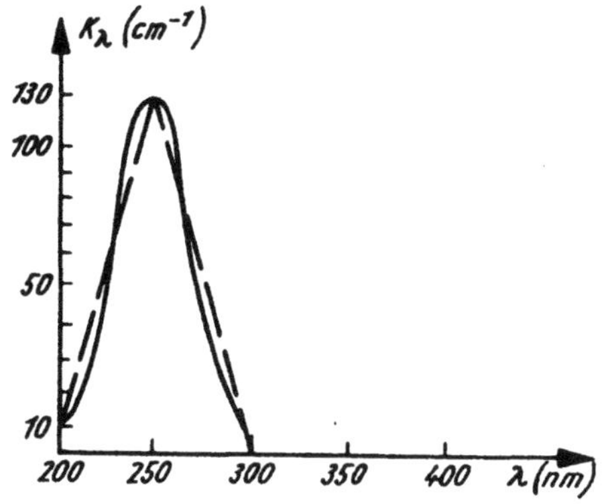

Fig. 13. Attenuation coefficient (K_λ) as a function of λ (for UV-radiation that passes the ozone layer)

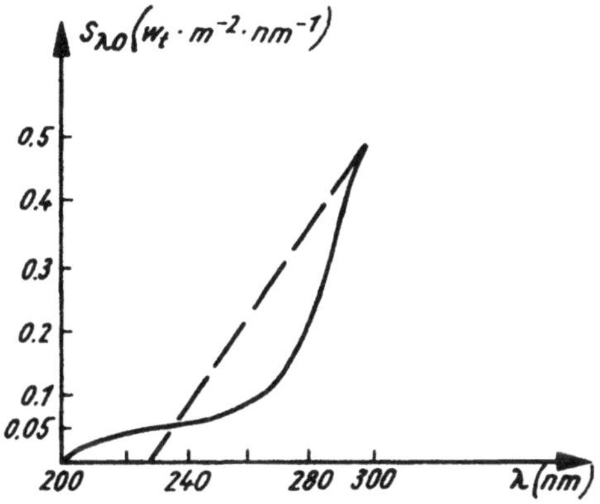

Fig. 14. UV-radiation flux at the upper stratosphere boundary as a function of wave length λ

If we integrate (1) S_λ by λ from 200 nm to 300 nm, considering that $h > 0.1$ cm and diminishing by terms of the order 10^{-130h}, we find the increase in intensity of UV-radiation:

$$\delta = \frac{S}{S_0} = \frac{h_0}{h} \, 10^{10(h_0 - h)} \tag{4}$$

A graph of the dependence of δ on h is depicted in Figure 15.

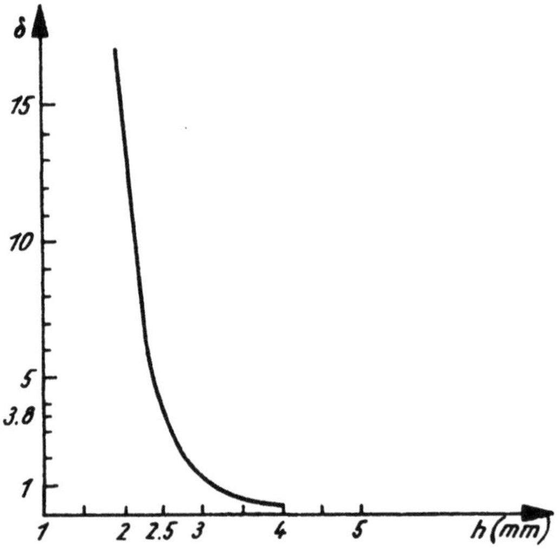

Fig. 15. Dependence of a relatively total UV-radiation flux on an amount of ozone in the atmosphere. Amount of ozone is measured by a thickness of ozone layer for normal temperature and pressure on a level of the surface of the earth

Let us calculate the increase in UV-radiation for a decrease in h to 0.1875 cm (h_0 = 3 cm). Using (4), we find that UV-radiation in this case increases 21 times. Obviously this is the maximum estimation.

Let us now estimate the thickness of the ozone layer and thus the level of UV-radiation one year after a nuclear conflict. This corresponds to the end of the "nuclear winter", when the transparency of the atmosphere would be restored. We will consider the rate of ozone formation (q) in the atmosphere to be constant. The characteristic residence time of ozone is several days in the upper stratosphere and about a year in the lower stratosphere. We will assume that the residence time of ozone in the atmosphere is one year. We can write a differential equation describing the dynamics of ozone in the atmosphere in the following way:

$$\frac{dh}{dt} = q - mh,$$

where q is a rate of natural generation of ozone, m is a decay coefficient.

According to [39], m = 1 $year^{-1}$. It is then easy to calculate that while the thickness of the ozone layer immediately after a nuclear conflict would be decreased to h = 0.1875 cm, after one year this thickness would recover up to a value of h_1 = 0.86 h_0; UV-radiation would thus be 3 times higher than at the beginning of a nuclear war. This level is comparable to the level of UV-radiation currently found in high mountainous regions.

6.2. The effect of UV-radiation on the biota

There exists a good deal of research studying the impact of UV-radiation on living systems [23, 24, 53, 102]. It is clear that even maximum estimations of the increase in the level of UV-radiation (and these are quite improbable) give values well below lethal doses (see Figure 16). It is therefore necessary to analyze the sub-lethal effects. These effects are manifested in the increased danger of irradiation by wave lengths of 240 nm to 290 nm for nucleic acids, enzyme proteins

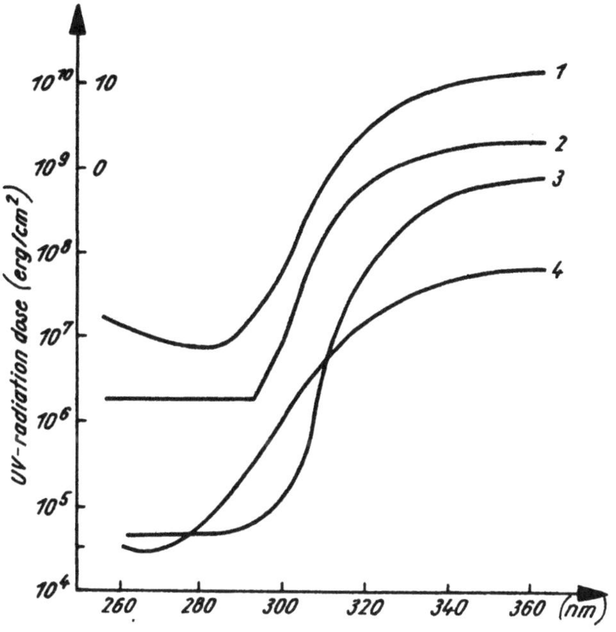

Fig. 16. Spectrum of lethal doses of UV-radiation for various organisms [24].
1 – plants (pollen); 2 – protozoa (Paramecium); 3 – insects (Drosophila);
4 – bacteria (E. coli)

and polipeptides. It is well-known that UV-radiation suppresses processes of cell repair. Therefore the synergistic effect of "radiation + UV-radiation" can be dangerous, since processes of post-radiation repair would be suppressed.

However, all these effects can come into play when the intensity of UV-radiation is increased 5–10 times, whereas we obtained a maximum estimation of 3 times. Much remains unclear, making further research necessary. The first question: how do ozone and oxides of nitrogen interact in the lower stratosphere? We used a very simple model, but it is unclear to what extent this model is commensurate with reality. There are many other questions as well.

UV-radiation is one of the most important ecological factors. UV-rays possess an extremely high reactivity (i.e. capability to bring about the lethal effect), which significantly exceeds the efficiency of visible-spectrum rays and cannot be explained by the difference in the energy levels of the quanta alone. Rather the existence of a great number of organic compounds absorbing actively the UV-rays is a basis for a wide variety of responses of living organisms to UV-radiation.

On the one hand, UV-radiation has the features characteristic of the ionizing radiation effects (viz. chromosome aberration, the mutagenic effect, changes in metabolism) and, on the other hand, it has the features similar to visible-light action (acceleration of flowering, growth stimulation, pigment-formation).

In a study of UV-radiation effects, attention must be paid to the spectral composition, the intensity and the dose capacity of UV-rays. In the UV-radiation impact on plants, the strong dependence of an outcome upon a spectral composition of UV-rays is observed. The perniciousness of UV-rays increases sharply with a decrease in wave length.

The natural boundary of the UV-radiation wave length on the surface is $\lambda > 290$–300 nm [101]. The least recorded wave length of UV-rays that ever reached the surface of the earth is 283 nm

[23]. The rays of a wave length of about 295 nm are of particular biological interest as the strong energy absorption by nucleic acids, aromatic amino acids, and peptide links is observed for this wave length. Under high doses, the rays of a wave length of 290–320 nm heavily damage the plant leaves, leading to weakening of the plants and decrease in their productivity [90].

As it was shown for various kinds of plants, an eventual effect (such as an accumulation of dry weight, vegetative biomass, or total leaf area) depends upon the total dose of UV-radiation absorbed by the plant.

An irradiation effect also depends upon intensity of UV-light and hence, under the same doses, upon the time during which the irradiation takes place. The dose of 60 $W \cdot m^{-2} \cdot min$ ($E = 1\ W/m^2$, $T = 60$ min) inhibits the growth of corn while the same dose 60 $W \cdot m^{-2} \cdot min$ ($E = 10\ mW/m^2$, $T =$ 10 hours) exerts a stimulating influence.

The influence of a long-wave UV-irradiation ($\lambda = 330$–390 nm) is of low efficiency under a short action, but is efficient when the irradiation is prolonged and of high intensity (7 W/m^2); for example, the dry weight of cucumbers and corn under this irradiation was 20–30 % higher than under normal conditions.

In a study of lethal effects of UV-rays on cells, the spectrum range of survival was found to be very similar for the cells of different organisms: the lethal effect sharply increases leading with 290 nm [24].

The same situation is observed for both the suppression of photosynthesis and inactivation of cytoplasm motion. The universal form of the UV-effect spectral curves for the death of organisms and for the suppression of their physiological properties seems to be an indication to the fact that the effect acceptors are common to all these cases (viz. the protein and nucleic components of cells).

In the course of evolution plants have developed certain systems of photoreactivation defence from UV-damage. Photoreactivation recovery is observed up to a certain threshold level of UV-radiation doses, beyond which the photoreactivation level rapidly decreases.

The threshold doses, which characterize the UV-resistance of various groups of plants, are given in Table 1.

Table 1. Classification of plants according to the threshold dose of UV-radiation [83] ($\lambda = 254$ nm), $10^5\ J/m^2$

Highly sensitive	0.4 – 2
Sensitive	2 – 10
Moderately sensitive	10 – 42
Moderately resistant	42 – 170
Resistant	170 – 500
Highly resistant	500 – 1250

In the cited research, where 67 different plant species were investigated, the range of their UV-resistance was found to be very wide. Plants with broad horizontal leaves are most sensitive (e.g. pea, kidney bean, etc.). Cereals (corn, wheat) are more resistant than most broad-leaved herbaceous plants. Succulents and some conifer species were highly resistant to UV-irradiation. Young specimen of agaves perished only under the UV-radiation that reached 3000 cal/cm^2 while pine needles died off under the dose of 2800 cal/cm^2, which is thousand times more than the highest lethal doses for living cells.

Dependences of the lethal UV-radiation doses on wave length for various organisms are shown in Figure 16 [24]. It can be seen that the less the UV-radiation wave length, the higher is the reactivity of the rays.

An average annual level of the solar radiation falling onto a unit area of the surface of the earth amounts to 140–300 W/m^2. Since UV-radiation accounts for about 1 % of the solar spectrum and the increase in UV-radiation in the year after the nuclear conflict (i.e. cessation of the "nuclear winter") would be three times the pre-conflict level, we obtain that the UV-radiation intensity near the surface would be 4–9 W/m^2. Comparing this quantity with the damaging and lethal levels of the UV-radiation affecting plants, we find that the UV-radiation would be pernicious for some species while for others there would be no obvious disturbances.

For example, the UV-irradiation under which a plant of a long light period is completely damaged amounts to 0.25–0.6 W/m^2 [23], which is 15 times less than the above-considered intensity of radiation.

Inhibition of photosynthesis in the sedges from foothills begins under a UV-radiation intensity of 5 W/m^2 while in the Alpine sedges it begins under 14 W/m^2. Hence, in the former case, the UV-radiation would result in the inhibition of photosynthesis, but in the latter case it would not.

6.3. Conclusion

After a nuclear conflict, the UV-radiation would increase as a result of the destruction of the ozone layer due to the release of nitrogen oxides produced by numerous fires into the atmosphere. The effect of nuclear detonations on the ozone layer thickness would be insignificant. The recovery of the ozone layer would take 5–8 years. A number of plant species would suffer inhibition of photosynthesis while other species would be able to endure the increased level of UV-radiation. On the whole, the photosynthesis in terrestrial plants would be depressed. The immune system would be partially disturbed in man and animals.

It is known [24] that even a slight increase in UV-radiation results in the depression of photosynthesis in phytoplankton. Since phytoplankton is a basis for the trophic chain of aquatic ecosystems, the increase of UV-radiation might lead to their degradation. In any case, the ecosystem of the open ocean, where the "phytoplankton" trophic chain dominates, would be most vulnerable even to a slight increase in UV-radiation.

It should be noted that the considerations of Subchapter 1 give the maximal estimations of the increase in UV-radiation, which result from the assumption that all the nitrogen dioxide generated by nuclear explosions and consequent fires would be emitted into the stratosphere and would combine with ozone. In reality, this amount would naturally be lower, which would lead, in turn, to the lower levels of UV-radiation. The effect of the "increase in UV-radiation" factor would be apparently not so important (compared to the effects of other factors).

7. THE EVOLUTION OF ECOSYSTEMS AFTER A NUCLEAR WAR "NUCLEAR SUCCESSIONS"

7.1. Introduction

How would these factors affect ecosystems, how would the natural equilibrium be distrubed, in what direction would the evolution of ecosystems proceed, where would all this lead to? How would global biogeochemical cycles change? Let us try to answer these questions, using the biogeographical principle of consideration. We will use general biological laws, as well as mathematical modelling.

7.2. Coniferous forests in the northern hemisphere

20–80 % of the coniferous forests of the northern hemisphere would die out as a result of the direct impact of nuclear explosions and subsequent fires (see Chapter 2). The surviving trees would be subjected to the effects of the "nuclear winter". The season of the year in which a nuclear conflict begins is key. If it began in winter, approximately 30–60 % of all trees might survive a year of "nuclear winter" (see Chapter 4). If it began in summer, the decrease in illumination would lead to the cessation of photosynthesis, and the cold would lead to the freezing of cellular water and the destruction of cells. As a result, almost all trees would die. However, in both cases seeds capable of germination would be preserved in the soil.

After the end of the "nuclear winter" other factors that might lead to the death of the biome of coniferous forests, would begin to take effect:

a) formation of "windows" in the tree canopy as a result of explosions and fires, the burning off of the upper layer over large areas, UV-radiation, and an increase in the concentration of CO and other pyrotoxins in the atmosphere would all lead to the suppression of photosynthesis, the subsequent decrease in transpiration and an increase in the accumulation of water in the soil;

b) acidification of the soil by "acid rain", which is formed when oxides of nitrogen and sulphur are washed out of the atmosphere, the contamination of the soil by radioactive substances and heavy metals would lead to the weakening of root systems, the intensive leaching of nutrients from the soil and, as a result, to a decrease in photosynthesis and an increase in the accumulation of water in the soil;

c) under the impact of "acid rain", radioactive and heavy metal contamination, and UV-radiation, the microflora and fauna of the surface layer of the soil would be suppressed, and the rate of decomposition of litter would be decreased; as a result the litter would absorb more water;

d) the appearance of massive outbreaks of forest pests would occur both as a result of the weakening of the defences of trees (the impact of "radiation + UV-radiation"), and as a result of weakening the pressure by "predators", e.g. birds.

All these factors, operating according to the principle of positive feedback, would result in the "nuclear" succession "forest-wetland", the rapid pallustrification of the territory previously occupied by coniferous forests (in humid zones). The characteristic time period for this process is 100–200 years under natura conditions. The "nuclear" succession would occur several times faster. This process would proceed especially rapidly in northern regions, where two-component soil-forming rocks – sands and sand loams – bedded at 40–60 cm by loam or clay are widespread. This causes high levels of ground-water beneath forests (40–60 cm) and a correspondingly significant percentage (up to 70 %) of gley-podzol and boggy-podzol soils. Under these conditions the woody layer acts as a kind of pump, pumping out excess moisture and intercepting a significant amount of precipitation. Therefore its destruction would raise the input in the water balance by approximately 200 mm [36]. Consequently we can conclude that the destruction of the tree cover over such large areas would also raise the level of ground-water in the remaining forests. This would trigger the progressive pallustrification of more than half of the taiga forests at latitudes of 60–65° N in the European part of the USSR.

For the coniferous forests of North America, secondary (caused by fire) pallustrification would take place on a significantly lower scale, having an effect only on the narrow strip of taiga-frozen soils in the vicinity of Hudson Bay. However, even in this region the reduction of the area of forests would not be significant because of the sharpness of relief and the good drainage of upland areas [107].

However, this conclusion does not seem to hold for all systems of coniferous forests. This is due to the fact that precipitation falling on land is formed not only by evaporation from the ocean, but by evaporation from land as well. Therefore, as a result of the pallustrification of territories adjacent to the ocean, transpiration, and thus total evaporation from the region would be reduced. Consequently the transport of water vapor deep into the continents would be reduced and the aridization of continental ecosystems might take place. For a quantitative analysis it is necessary to calculate the variation in the distribution of precipitation over the continents.

7.3. Deciduous forests of the northern hemisphere

The same climatic factors would have a similar effect on deciduous forests: almost total death if the conflict took place in summer, and probable survival if it occured in winter.

Under normal conditions defoliation and transition to dormancy take about 1.5 months. Since the onset of the "nuclear night" would be practically instantaneous (several days) and the "nuclear winter" would begin within two weeks to a month, trees would not have enough time to make the transition to the winter stage and would die out. More deciduous trees than conifers would die (approximately 70—100 %) if the nuclear conflict were to take place during the winter. This is due to the lower resistance of deciduous trees to cold.

The trees which survive a "nuclear winter" and their sprouting seeds would be affected by the same factors as conifers. However, the "nuclear" succession of the "forest-grassland" type would play a more significant role. Forest ecosystems are rather quickly replaced by grasslands. In some (more well-preserved) areas this succession would not take place, but tree development would be suspended for many years, with renewal occurring more slowly than after forest fires.

7.4. Grasslands

The soil would not be significantly frozen, since it would be quickly covered by snow. The root systems of some perennial grasses, seeds, microflora (and to some extent microfauna) would survive. However, because of the abundance of ruderal seeds, absence of animals, and acidification and erosion of the soil, the "nuclear" succession would most probably proceed in the direction of ruderal ("garbage") vegetation. Ruderal and dwarf shrub vegetation would spread to the areas occupied by remainders of inhibited deciduous forests.

7.5. Tropical and subtropical rain and monsoon forests, savannah

These ecosystems would not survive the climatic stress. Absence of thick litter and extensive root systems (in forests) would lead to a rapid water and wind erosion of tropical and subtropical soils, exposure of bedrock, and leaching of nutrients, rendering these regions completely unproductive. Prolonged exposure of laterite soils (red soils), which are widespread at latitudes from 30° S to 30° N, would render them hard as mountain bedrock, and thus unsuitable for both agriculture and the restoration of natural vegetation. "Nuclear depression" of the savannah would lead to a reduction in total transpiration, invasion by thorny shrubs, and the formation of semidesert landscapes followed by rapid desertification of these territories.

The "nuclear successions" in land ecosystem mentioned above may be classified as consequences of a nuclear war, which, though they develop slowly (over dozens of years), radically change the biogeography of the planet.

On the whole, the degradation of vegetation, i.e., "nuclear" successions of the "forest-wetland", "forest-grassland" and "grassland-ruderal grassland" types, would occur throughout the northern hemisphere. About 30 % of plant species (according to expert estimations), and almost all species of mammals and birds would die out. Decomposition of dead organic material would practically cease in the course of the year following "nuclear winter", but would then be restored with the regeneration of soil flora and fauna. Severe soil erosion would occur. The restoration of ecosystems to their pre-conflict state would be impossible because of species extinction.

The above ecological consequences would affect the southern hemisphere as well, but to a lower degree, since climatic and radiation stress would be weaker there.

7.6. Use of modelling of global biospheric processes

The model of global biospheric processes developed at the USSR Academy of Sciences Computer Center [39] incorporates a description of ecological, climatic and demographic processes on the earth. The goal of the model is to analyze possible paths of the co-evolution of human civilization and the biosphere. The model was developed to analyze "peacetime" situations, and was not intended for use in an analysis of military events and their consequences. However, it is evident that with some supplementation the model can be used in calculations for some situations which might arise after a nuclear conflict. There are three main difficulties.

The first is the difficulty of determining the situation after a nuclear conflict (degree of degradation of natural systems, degree of destruction of demographic and economic potential). This means a certain indefiniteness in the initial data used in the calculations. The estimations made in this report serve to overcome this difficulty to a certain degree.

The second difficulty lies in the fact that we know very little about the ability of natural systems to renew themselves after suffering the complex impact of a nuclear conflict. On the one hand, research in the area of environmental protection, including the study of the effects of agricultural activity on natural systems, is carried out in the Soviet Union and other countries. Some of these quantitative results can be used (with certain reservations) in estimating the development of ecosystems after a nuclear conflict. On the other hand, there exists a relatively small number of reports that describe the renewal of vegetation on nuclear weapon test sites [112].

The third difficulty stems from the fact that we have to determine the socio-economic situation after a nuclear war, i.e., the probability that old social institutions would continue to function or that new ones would arise. This does not mean that we are ruling out the possibility of the extinction of Homo sapiens as a biological species. The probability of this occurring is quite high. In our calculations we will assume that the general structure of socio-economic institutions will remain intact, since we have no other hypothesis at present.

In modelling the consequences of a nuclear war, the "environment" bloc can be viewed independently, since the energy and technological capabilities of surviving man would be so low that the impact of man on the environment in the first few decades after a nuclear war would be insignificant. On the other hand, the functioning of "demography" and "human activity" blocs would be determined not only by their own initial states, but also by environmental factors.

7.7. Global biochemical cycles

The intensity of the cycle of chemical elements (carbon, nitrogen, etc.), as well as the quantity of matter taking part in the cycle would decrease for most ecosystems. As a result, the amount of CO_2 in the atmosphere would increase, as would the accumulation of biogenic elements in reservoirs. The amount of oxygen in the atmosphere would not be significantly reduced.

A great amount CO_2 would be released into the atmosphere from fires. The death of trees as a result of climatic and radiation stress would lead (on account of the decomposition of organic substances) to an additional flow of CO_2 into the atmosphere. As a result of the decrease in productivity of land plants, the amount of humus would also be reduced. Consequently, land would become a source of atmospheric CO_2. The absorption of excess atmospheric CO_2 would be determined by the ocean for a long time.

Let us estimate the change in the amount of CO_2 in the atmosphere and the change in average global temperatures using the following scenario. Assuming that 20 % of the forests of the northern hemisphere were completely burned at the time of the fires, we would find that the amount of CO_2 in the atmosphere is raised by 15 % almost instantaneously. Then, during the "nuclear winter" all forests of the northern hemisphere and tropics would die out. The corresponding areas would be overgrown with grass and dwarf shrubs within five years. Three years after the "nuclear winter", the processes of decomposition of dead organic material, litter and humus would be completely restored. The transparency of the atmosphere would be restored after the "nuclear winter".

The course of variations in atmospheric CO_2 and in average global temperatures was calculated according to this scenario, using the model of global biospheric processes [39] (see Figure 17). This scenario differs from the scenario described in Chapter 3 mainly in the amount of carbon emitted into the atmosphere as a result of fires. More exact calculations show that 2×10^{10} t of carbon would be instantly emitted into the atmosphere owing to forest fires and "fires of civilization", leading to an increase in the amount of CO_2 in the atmosphere of approximately 3 %. Here we have chosen a more "severe" scenario, considering that all organic matter would be burned. However, the ensuing CO_2 dynamics (and, of course, temperature) would be practically independent of initial emissions, since the flow of carbon into the atmosphere three years after the war would come from the decomposition of dead organic matter from plants that had died during the "nuclear winter".

In 30 years the amount of CO_2 in the atmosphere would be increased 1.6 times and the temperature (on account of the "greenhouse effect") would rise by 1.3 °C. This would be followed by a slight decline, which would last 100–150 years, until the end of this "nuclear" fluctuation.

The general conclusioh that can be drawn concerning biogeochemical cycles is that the elimination of forests and the replacement of forest ecosystems by mire and grassland ecosystems would sharply reduce the stability of the biosphere as a whole and its ability to damp climatic variations. This is due to the fact that forest ecosystems take the main part in the regulation of the carbon cycle and the related global atmospheric temperature. The climate would therefore become less stable.

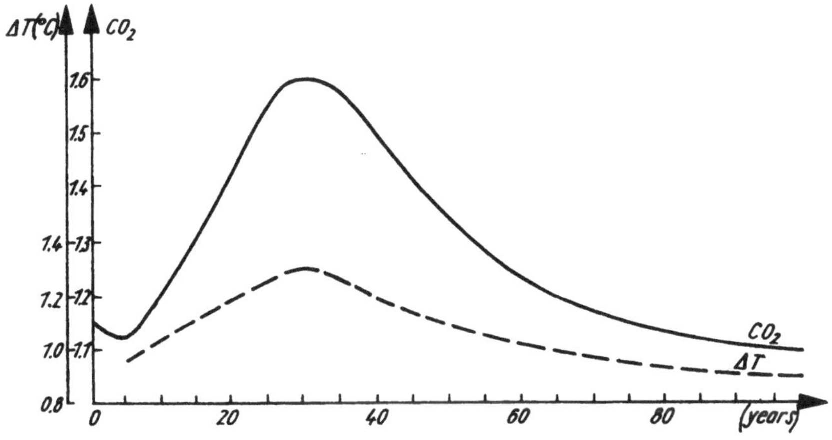

Fig. 17. Variations in CO_2 concentration in the atmosphere (in relative units with respect to the current concentration of 330 ppm). Dynamics of the deviation of the annual global temperature from the current value of 15 °C
0 – beginning of nuclear war

7.8. The process of the renewal of vegetation (modelling)

In order to obtain quantitative estimations on the renewal of vegetation in a zone damaged by factors of nuclear war, the model of global biospheric processes was supplemented by the bloc of the degradation and renewal of forest and grassland ecosystems [5]. The degradation of vegetation in forest ecosystems in the northern and central regions and tropics, as well as grassland ecosystems, is considered in this bloc.

In terms of external impact, we will consider pollution by complex pollutants leading to the suppression and cessation of photosynthesis, the retardation of the decomposition of litter and humus, and a sharp increase in washing them away due to the death of plant root systems (soil erosion).

The model describes the cycle of carbon and nitrogen in the ecosystem. Vegetation is divided into three types: grasslands, deciduous forests, coniferous forests. In the model, photosynthesis depends on the concentration of pollutants, the amount of nutrients in the soil and the type of vegetation. Pollution affects all the other processes that occur in an ecosystem: dying off of vegetation, formation and decomposition of humus, etc.

We indicate the results of modelling the renewal of ecosystems after a nuclear war for those cases in which initial vegetation is restored, i.e., where there is no "nuclear succession". Since the rate of renewal is largely dependent on the degree of soil erosion at the moment when renewal begins, we considered several variants. We used a scenario in which vegetation died off completely as a result of the effects of a nuclear war. Severe soil erosion occurred shortly after the nuclear war. The effect of pollution ("ordinary" and radioactive) was cut off. Vegetation began to renew itself.

Let us consider the variant of the renewal of a typical mixed forest in the central latitudes. The results of our calculations are contained in Fig. 18, where the initial moment corresponds to the cessation of pollution and the beginning of plant development. It is assumed that, as a result of soil erosion at the initial moment, 1 % of the initial amount of humus remained intact. It is evident that grass vegetation would begin to renew itself during the first few years, while trees would take much longer.

The accumulation of humus depends on the course of the renewal of grass vegetation. Renewal of deciduous and coniferous trees does not occur until the humus content reaches a certain threshold value. Therefore renewal of deciduous trees begins 20–30 years after the beginning of grass vegetation renewal, while the renewal of conifers begins after 70–80 years.

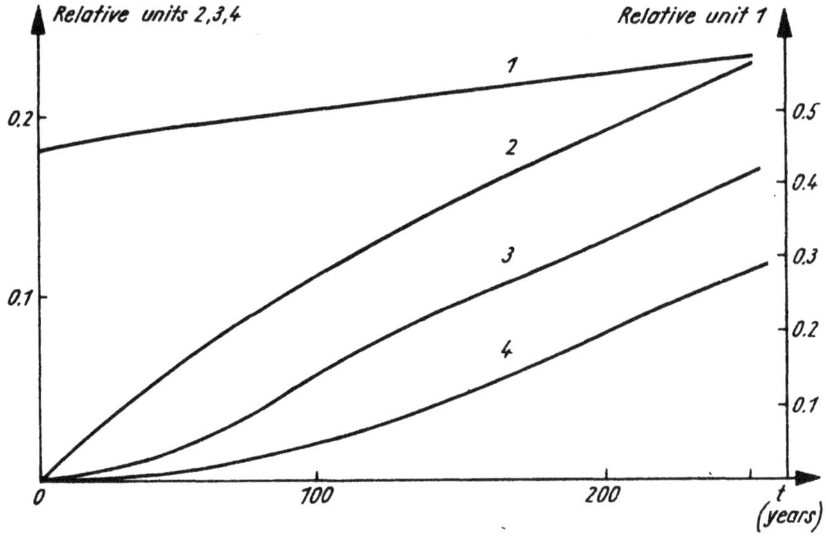

Fig. 18. Renewal of vegetation of mixed forests in the region of central latitudes which have suffered the effects of a nuclear war

1 – grass productivity
2 – humus amount
3 – biomass of deciduous trees
4 – biomass of coniferous trees

Variables are measured in relative units (in relation to values before the nuclear war)

The renewal of vegetation and humus occurs very slowly. 200 years after the beginning of the renewal, the biomass of deciduous plants is 13 % of the biomass before the nuclear conflict; for coniferous trees this value is 8 %. The amount of humus at this time reaches only 20 %, while the productivity of grass vegetation reaches 55 %. Complete renewal takes more than 2000 years.

Results of calculations for deciduous and coniferous forests in the central latitudes as a factor of the degree of soil erosion at the beginning of renewal are contained in Tables 1 and 2. We can see that the rate of the renewal of biomass and productivity is largely dependent on the degree of soil erosion. Renewal takes place most quickly when there is no erosion. In 50 years the biomass of a deciduous forest in this cycle reaches 71 % of its value before the nuclear war. The rate of renewal then slows down. A coniferous forest renews its biomass more slowly; in 50 years up to 41 % of the biomass is renewed.

If 50 % of all humus is preserved at the beginning of the renewal, in 50 years a deciduous forest renews up to 33 % of its biomass, while a coniferous forest renews up to 18 %. It is essential to note that after the beginning of the renewal of vegetation, the reduction in the amount of humus and litter continues for several years. This is related to the fact that plant root systems are not immediately renewed. The washing away of humus and litter is cut off as the renewal of root systems occurs. Furthermore, during the first few years of renewal, productivity is low and the dying off and decomposition of litter-fall cannot compensate for the decomposition of humus and litter. If 50 % of all humus and litter is preserved at the beginning of renewal, the reduction in the mass of litter continues for another 5 years, and the minimum amount of humus will be reached in the 25-th year. The minimum values for humus and litter are 26 % and 42 %, respectively.

Table 1. Recovery of a deciduous forest in a temperate zone as a function of a soil erosion degree in the beginning of the recovery process. Presented are relative values of tree biomass (B) and tree productivity (P) (relative to the values before the war) in various years after the beginning of the recovery process

Degree of humus disruption (%)	25 years		50 years		100 years		200 years	
	B	P	B	P	B	P	B	P
0	0.5	0.83	0.71	0.83	0.82	0.85	0.82	0.85
25	0.36	0.61	0.51	0.61	0.6	0.61	0.61	0.62
50	0.22	0.36	0.33	0.44	0.45	0.48	0.49	0.5
75	0.1	0.16	0.14	0.2	0.27	0.35	0.45	0.4

Table 2. The same for a coniferous forest

Degree of humus disruption (%)	25 years		50 years		100 years		200 years	
	B	P	B	P	B	P	B	P
0	0.24	0.8	0.41	0.8	0.62	0.8	0.77	0.81
25	0.17	0.56	0.3	0.57	0.44	0.58	0.56	0.6
50	0.1	0.32	0.18	0.4	0.32	0.45	0.43	0.48
75	0.04	0.13	0.08	0.17	0.17	0.32	0.37	0.49

In grass ecosystems of the chernozem zone there will be a significant reduction in the amount of humus as a result of soil erosion. However, because of greater warmth and a lower dependence on the amount of humus, the productivity of these ecosystems will be higher than in forest ecosystems of the temperate zone. This causes more intensive growth in grasses, but the renewal of humus will occur more slowly. This is related to the fact that grass systems accumulate much more humus than forest systems. In simulation experiments it has been shown that in cases where 1 % of all humus remains at the beginning of renewal, it takes 200–300 years for up to 10 % of the original amount of humus to be restored.

It should be emphasized that the death of vegetation in tropical forests is possible only as a result of the effect of the "nuclear winter". Because of the complete death of plants and the denudation of soil, erosion in these ecosystems would occur very rapidly and intensively.

The denudation of laterite soils (red soils) leads to their complete non-fertility (see Section 7.5.). This all gives us a basis for concluding that the complete renewal of vegetation of tropical forests is impossible.

Calculations in the model for tropical forests (deciduous trees and grasses) show that after the factors of a nuclear war cease to have effect, only grass vegetation with a productivity of approximately 5 % of the original productivity will grow. The accumulation of humus begins to take place so slowly that it takes 300–400 years for the conditions necessary for tree growth to appear. Thus the model shows that the renewal of vegetation and soil in tropical forests will take place extremely slowly.

If we now compare the characteristic time period for the renewal of vegetation in various geographical zones after a nuclear war, we find that ecosystems in the equatorial zone and ecosystems of northern forests are renewed most slowly. Although ecosystems in the temperate zone are renewed more quickly, even here the rate of renewal is extremely low.

7.9. The geographical distribution of the process of vegetation renewal

In order to estimate how vegetation would be renewed in various regions of the planet we used our model for calculations according to the following scenario:

90 % of the forests in Northern America and Eurasia above 50° N latitude were burned; after one year grassland and wetland vegetative communities begin to form in place of the forests. In the remaining types of vegetative communities, 20 % of the terrestrial organic material was destroyed by fire; one year later these communities begin to renew without changes in type.

The process of renewal is patchy with the minimum of renewal occurring in areas formerly covered by forests. Figure 19 contains the results of calculations showing how and where the productivity of vegetative communities renews itself after 15 years. It is interesting that this factor leads to changes in productivity not only in the places where it took effect, but also at other points on the planet. For example, it leads to a significant decrease in the productivity of vegetative communities in the Amazon watershed and in Southeast Asia.

It should be noted that in this scenario we have not taken the "nuclear winter" or the sharply increased level of radioactive contamination into account. In fact, this scenario shows only the consequences of a massive dying out of vegetation in the northern hemisphere. However, as we can see from the calculations, this alone is sufficient to cause a significant decrease in the productivity of vegetative communities on the planet.

7.10. The problem of reduction in the diversity of species and the stability of the biosphere

Obviously climatic and radiation stress, as well as fires, would lead to a sharp and rapid reduction in the diversity of species. The characteristic time period for this process is on the order of several months. However, as a result of nuclear war there arises still another factor. The effect of this factor will be prolonged over many years and decades, but the result will be the same – the acceleration of the extinction of species and a reduction in the diversity of species. This factor is the insularization (from the Latin "insula" – island) of territories.

Clearly, as a result of explosions the initial almost continuous distribution of biota over the

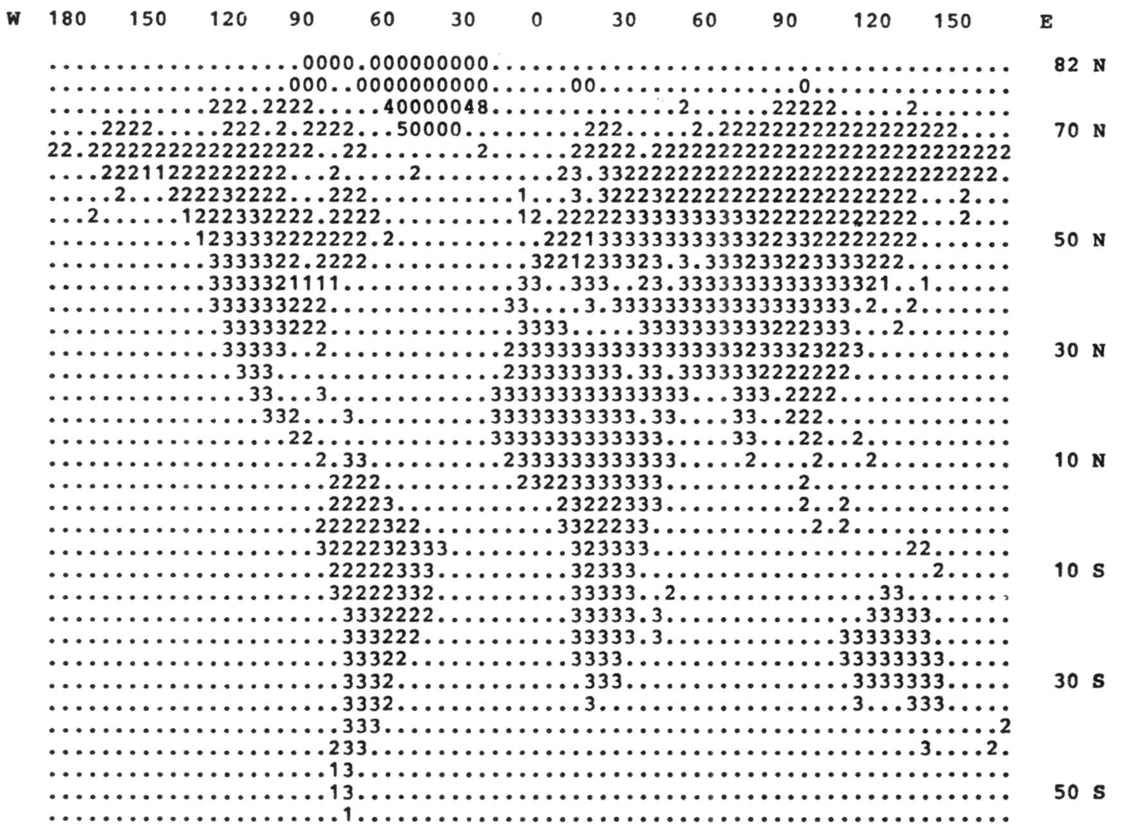

Fig. 19. Computer map which shows values of productivity of terrestrial ecosystems 15 years after the nuclear war. Regions in which the productivity is 15 % less than the productivity before the beginning of nuclear war are marked by 1, 10 % less than before are marked by 2. Regions in which the productivity is practically previous, are marked by 3 and 4. Figures of 5 to 9 show the regions in which the productivity increases more than 1 %

surface of the earth would be replaced by discrete distribution: territories of survival would alternate with territories almost completely devoid of life. In this case a significant reduction in the number of species, and hence a reduction in their diversity, would be caused not so much by a decrease in the area occupied by biota as by its fragmentation into small unconnected areals.

Let us consider several general questions in connection with the consequences of a nuclear war for the biosphere as a whole. The main question here is the question of the stability of the current state of the biosphere. For its macrodescription we have used the entropy measure, i.e., the information entropy that measures the diversity of species and ecosystems in the biosphere. (In ecology this has become an almost axiomatic measure of the stability of an ecosystem.) According to this measure, the stability of the biosphere will rise with an increase in the number of different ecosystems and an increase in the diversity of species. Obviously, as a result of a nuclear war both the absolute number of ecosystems and their diversity of species would decrease; these two processes lead to a reduction in the entropy and hence to a reduction in the global stability of the biosphere.

Let N be equal to the total biomass of life on the planet and let S be equal to its information entropy. It is found that if

$$dS/dt > dN/dt, \qquad (*)$$

then the stability of the biosphere will not be disturbed [39]. Even if we assume that a nuclear war and its consequences do not significantly reduce the total biomass of the biosphere, a sharp and significant reduction in the diversity of species and ecosystems would most likely take place, nevertheless. All this can lead (if the inequality (*) is distrubed) to the destruction of the current state of the biosphere and a transition to a new state, in which the existence of an ecological niche for Homo sapiens would be more and more problematic.

7.11. Conclusion

Factors of nuclear war would trigger the starting mechanisms of ecosystem evolution, leading to a change in the whole biogeography of the planet. "Nuclear successions" of the type we usually call "degradation" of ecosystems and landscapes (pallustrification, desertification, growth of ruderal vegetation) would take place. The denudation of soil as a result of the elimination of vegetation would lead to intensive and rapid erosion. As a result, the renewal of vegetation would occur at extremely low rates, and in some cases (tropical forests) renewal would be problematic at all.

The elimination of vegetation over large territories, the appearance of a large quantity of dead organic matter, and the replacement of forest ecosystems by grassland ecosystems would bring about serious disturbances in the global biogeochemical cycle of carbon, hence leading to climatic changes.

The sharp decrease in the diversity of species and ecosystems would reduce biosphere stability. It is quite possible that the current state of the biosphere would become unstable, with evolution beginning towards a new state. The issue of whether an ecological niche for the species Homo sapiens would exist in this state of the biosphere is open to question.

8. THE IMPACT OF FACTORS OF NUCLEAR WAR ON FRESH WATER ECOSYSTEMS

8.1. Introduction

As a result of a nuclear conflict, fresh water ecosystems would suffer the synergistic impact of radioactive contamination on the one hand and pollution by heavy metals on the other. We will devote most of our attention to water bodies: lakes, reservoirs, etc., rather than flowing water, due to the following reasons.

Firstly, the calculations we use for radionuclide and heavy metal load are most well-documented for these ecosystems. These relatively isolated ecosystems are quite inert (characteristic residence time for water is on the order of a year), in contrast to rivers, in which self-cleaning occurs quickly and processes are of a transitional nature.

Secondly, the situation for rivers depends to a much greater extent on its specific characteristics, the characteristics of the water shed and even on the scenario of precipitation in the period after a nuclear conflict. The same is true for ground-water, which, generally speaking, is of interest as the potentially least polluted source of drinking water. However, for ground-water located near the surface this does not hold true: for example, the concentration of radioactive contamination at a depth of approximately 1 m may be higher than on the surface [78]. The extraction of water located at great depths is substantially complicated by the absence of technical services and the disturbance of the infrastructure of the economy. Therefore, our estimations for fresh water reservoirs will be the worst in their own way. The situation in rivers would fluctuate within these estimates. Although it would soon become more favorable, the situation could sharply become worse at any time (for example, after recurrent rain), necessitating specialized analyses for every water use over the course of a rather extended period of time.

In order to get an idea of the impact on a certain abstract body of water we will use the concept of a standard water body system, with characteristics typical of lakes or reservoirs used for drinking water supplies. These are, as a rule, artificial or natural reservoirs located near cities or industrial centers (for example, the Kliazma reservoir near Moscow, the reservoir in Central Park in New York City, etc.). For these reservoirs it is important to fix two parameters:

1. an average depth h of 4 m;
2. the area of the water shed is 100 times greater than the water surface.

Generally speaking, the remaining parameters may be arbitrary; they do not affect our estimations. It should be noted that hypotheses 1. and 2. are fulfilled for most shallow water ecosystems used as major supplies of drinking water.

First we will estimate the concentrations of radioactive contamination in bodies of water for various periods of time after a nuclear conflict. Using this as a basis, we will then calculate dose loads, for which calculations will be made according to a specialized model taking into account processes of sedimentation and accumulation in sediment. Estimations will then be given for the release of heavy metals into fresh water bodies. All this will allow us to get a general picture of the state of fresh water ecosystems after a nuclear war.

8.2. Calculation of concentrations of radioactive contamination in fresh water ecosystems

8.2.1. Standard water body system

We will use data from Chapter 1 on concentrations of radioactive contamination of different points on the planet after a nuclear conflict. By placing the standard water body system in various zones of contamination, we can estimate the condition of water for each zone.

Let us first list the hypotheses on which we will base our estimations. We assume that radioactive contamination would be uniformly distributed over the territory of the watershed, and that an amount of radionuclides equal to their average concentration multiplied by the water surface area would be released directly into the water body system. In addition to the direct release of radionuclides, washing of radionuclides from the surface of the water shed in the moments that follow should also be taken into account. According to Wetzel's data [78], surface flow would be approximately 20 % of the total flow. If we use Joergensen's data [100] on the washing of pesticides by surface flow, by analogy we find that from 0 % to 40 % of the radionuclides would be leached. Therefore let us assume that 10 % of the products of fission released on the surface of the water shed would be washed into the standard water body along with surface flow, and that this would occur relatively rapidly – in the course of 20 days. The probability of rather excessive precipitation during this time is rather high, since as a rule, nuclear explosions are accompanied by rainfall [22]. For simplicity the age of the mixture of radionuclides upon entering the water body system will be averaged and set at 10 days. Ground-water would account for the remaining amount of radioactive contamination entering the water body from the water shed. According to [78], ^{89}Sr and ^{90}Sr are leached at a rate of 1.5 % per year, and ^{137}Cs is leached at a rate of 0.5 per year; ^{131}I is easily absorbed by humus and has a short half-life (8 days). Thus the radioactivity of ground-water would come only from ^{106}Ru, which would be gradually washed into the water body system. Taking into account the already practically pure surface runoff, we can assume that further leaching of fission products would not change the concentrations of radioactive contamination in the water body system, i.e., the concentration of radioactive contamination in drainage water would remain at the same level as in the water body system. In addition, it is assumed that in one day the radioactive contamination would be distributed uniformly over the entire 4-meter depth of the water body. Therefore, radionuclides falling to the surface with a total radioactivity A (Ci/km²) would occur in the water body in concentrations of

$$C = \frac{A}{h \times 10^9} = \frac{A}{4 \times 10^9} \quad \text{(Ci/l)}.$$

The ensuing decay of fission products and their redistribution among components of the water body (transition to bottom sediments) leads to a reduction of the concentration of radionuclides in the water (Table 1). Table 1 was calculated using a model describing the migration of radionuclides in a fresh water body [60]. The following formula was used for the calculation of the change in concentration of radionuclides in water in a single case of contamination:

$$u = u_0 \times e^{y^2} \, \mathrm{Erfc}(y),$$

where

$$y = \frac{K_d\sqrt{Dt}}{h}, \quad \mathrm{Erfc}(y) = \frac{2}{\sqrt{\pi}} \int_y^{\infty} e^{-y^2} dy,$$

u = concentration of radionuclides in the water at time t;
u_0 = the same for the initial moment in time (immediately following contamination);
K_d = the distribution coefficient for radionuclides between the solid and liquid phases in the water body;
D = the effective coefficient of migration of radionuclides in bottom sediment;
h = average depth of the water body.

In estimating the role of self-cleaning in the general change in concentration of radionuclides in water, we have used the values K_d = 800, D = 2 cm²/yr; these values were selected on the basis of averaging and analyzing the data contained in reports [47, 51, 62, 63]. The data presented in Table 2 indicate a rather substantial impact of radioactive contamination, since even in a zone of moderate contamination, over the course of a significant period of time, levels of contamination for water bodies exceed levels of concentration assumed to have "no apparent impact" on all groups of hydrobionts (Table 2).

If we compare the data contained in Table 1 with the maximum acceptable concentration of contaminating substances for the USSR (approximately 1–0.1 X 10^{-8} Ci/dm³ for most radioactive substances [56a]), we can see that the calculated levels of contamination are such that the use of water from open water bodies for the needs of the population even 5 years after a nuclear conflict would hardly be possible. The probability of being able to use various technologies for purification would be limited, and obviously rather ineffective as a result of the destruction of basic productive capacities.

Table 1. Radionuclide concentrations in water after a nuclear war (for the North American region)

Time since the outbreak of war	Concentration in water (Ci/l)								
	Zone of intensive contamination (e.g. eastern coast of the USA)			Zone of moderate contamination (south-east of the USA)			Zone of weak contamination (central regions of the USA)		
	A	B	C	A	B	C	A	B	C
24 hours	$8 \cdot 10^{-4}$	$2.6 \cdot 10^{-5}$	$8.3 \cdot 10^{-4}$	$1.6 \cdot 10^{-4}$	$5.2 \cdot 10^{-6}$	$1.7 \cdot 10^{-4}$	$5 \cdot 10^{-5}$	$1.7 \cdot 10^{-6}$	$5.2 \cdot 10^{-5}$
10 days	$3.9 \cdot 10^{-4}$	$1.0 \cdot 10^{-4}$	$4.9 \cdot 10^{-4}$	$8.0 \cdot 10^{-}$[illegible]	$2.1 \cdot 10^{-6}$	$1 \cdot 10^{-6}$	$2.4 \cdot 10^{-5}$	$6.5 \cdot 10^{-6}$	$3.1 \cdot 10^{-5}$
70 days	$1.7 \cdot 10^{-5}$	$2.4 \cdot 10^{-5}$	$4.1 \cdot 10^{-5}$	$3.4 \cdot 10^{-6}$	$4.8 \cdot 10^{-6}$	$8.2 \cdot 10^{-6}$	$1.0 \cdot 10^{-6}$	$1.5 \cdot 10^{-6}$	$2.5 \cdot 10^{-6}$
1 year	$1.2 \cdot 10^{-6}$	$3.7 \cdot 10^{-6}$	$4.9 \cdot 10^{-6}$	$2.4 \cdot 10^{-7}$	$7.0 \cdot 10^{-7}$	$9.4 \cdot 10^{-7}$	$7.5 \cdot 10^{-8}$	$2.3 \cdot 10^{-7}$	$3.1 \cdot 10^{-7}$
5 years	$1.1 \cdot 10^{-8}$	$2.9 \cdot 10^{-7}$	$3.0 \cdot 10^{-7}$	$2.3 \cdot 10^{-9}$	$6.0 \cdot 10^{-8}$	$6.2 \cdot 10^{-8}$	$7.0 \cdot 10^{-10}$	$2.4 \cdot 10^{-8}$	$2.4 \cdot 10^{-8}$

A – water contamination due to nuclear blast
B – water contamination due to destruction of APPs
C – total contamination

Note: For the USSR water contamination from blast would be approximately the same; contamination from destructed APP would be approximately 4 times lower

Table 2. Concentrations of radionuclides having no apparent impact (according to data contained in [89])

Groups of organisms	Acceptable concentrations of radionuclides (Ci/l)
fish	2×10^{-10}
crustaceans	1×10^{-6}
algae	5×10^{-5}
bacteria	5×10^{-5}
mollusks	5×10^{-5}

8.2.2. Onezhskoye Lake

Onezhskoye Lake has a volume of $V = 2.95 \times 10^{14}$ l, a surface area of 9890 km², a maximal depth of 110 m, and an average depth of 30 m. The lake is located in a zone of radioactive contomination from explosions in the cities of Petrozavodsk (total yield of 1 Mt) and Leningrad (10 Mt), and on the Leningrad nuclear power plant (4 Gw(e)). We will assume that a mixture of fission products will enter the water body in approximately 2 hours; as a result the surface layer of the water body (0.5 m) would contain 6×10^{-3} Ci/l. A cloud of radioactive contamination from explosions of nuclear weapons in Leningrad (total yield of 10 Mt) would reach the center of the water body in 14 hours, leading to an increase in the contamination of the water body of approximately 2.8×10^{-4} Ci/l calculated for a surface layer of 2 meters. Contamination of the water body by radioactive products from the destruction of nuclear reactors would begin in about 15 hours. As a result, after 16 hours the concentration of radionuclides in a 2-meter surface layer would be approximately 4.2×10^{-4} Ci/l. Products from the nuclear reactors of the Leningrad nuclear power plant would make up about 12 % of the total number of radionuclides entering the water body (see Fig. 20).

After one year the products of fission would be distributed throughout the entire depth of the

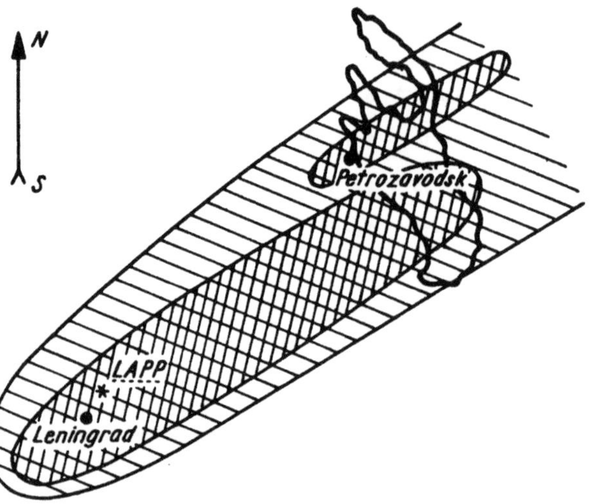

Fig. 20. Distribution of radioactive contamination along the Onega lake

● – cities with population more than 1 million
○ – cities with population less than 1 million
* – APP (atomic power plants)
///// – radioactive contamination zones which were formed by explosions in cities
\\\\\ – radioactive contamination zones which were formed by destruction of atomic power plants

water body, leading to a decrease in the average concentration of radionuclides in the water body to approximately 1×10^{-7} Ci/l, which would stem almost completely from radioactive products from the Leningrad nuclear power plant. In 5 years the concentration of radionuclides in the water body would be decreased approximately 10 times and would represent a value on the order of 1×10^{-8} Ci/l, i.e., below the maximum acceptable level.

8.2.3. Lake Ontario

Lake Ontario has an area of 19,500 km², a volume of 1710 km³ and an average depth of 88 m. The lake might suffer from contamination from explosions with a total yield of 10 Mt (Detroit), two bombs of one megaton each, and the destruction of 3 nuclear power plants with an average power (according to our estimations) of 6 Gw(e) each. Contamination would vary depending on winds.

First scenario: west wind. Contamination from the explosion of 10 Mt and two nuclear power plants can be considered to correspond to the percentages of contaminated area shown in Figure 21a. A cloud of contamination from the city and more distant nuclear power plants would reach the center of the lake in 19 hours, leading to an increase in the concentration of radionuclides in the upper 2 meters of the lake to 1.2×10^{-3} Ci/l. Contamination from the first nuclear power plant, which would reach the lake first, would create a concentration of 1.3×10^{-2} Ci/l in the upper 0.5 meters. As a result, the concentration in a 2-meter layer at the end of the first day would be 0.5×10^{-3} Ci/l. Taking into account the dispersal of contamination throughout the

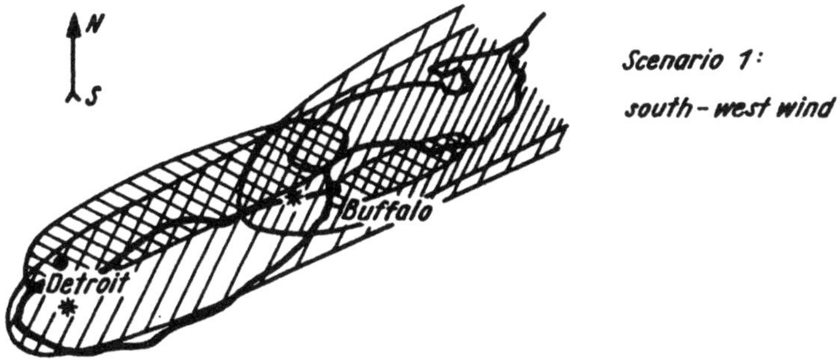

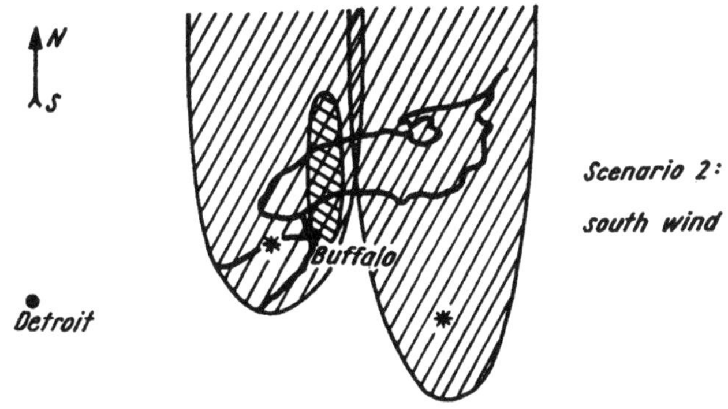

Fig. 21. Distribution of radioactive contamination (RAC) along the Ontario lake for two scenarios

● – cities with population more than 1 million
○ – cities with population less than 1 million
* – APP (atomic power plants)
///// – radioactive contamination zones which were formed by explosions in cities
\\\\\ – radioactive contamination zones which were formed by destruction of atomic power plants

entire depth of the water body and inflow into the water body (residence time in the lake is 10 years; inflow is supplied by relatively uncontaminated water from the other Great Lakes), the concentration after one year would be 8.55×10^{-7} Ci/l. Since a substantial contribution to contamination would be made by the mixture of isotopes from the destruction of nuclear power plants, which decay significantly more slowly than isotopes from a nuclear explosion, it would take 5 years for the concentration of radioactive contamination to reach a level of 0.52×10^{-8}, i.e., below the maximum acceptable level.

Second scenario: south wind (Figure 21b). Contamination from explosions of 2×1 Mt and two nuclear power plants. In this case radioactive products would enter the lake more quickly and the initial concentration in a 0.5 m layer would be 2.6×10^{-2} Ci/l; after one day the concentration in a 2 m layer would be 1.1×10^{-3} Ci/l; after one year the concentration of radioactive contamination throughout the entire water body would be equal to 3.2×10^{-6} Ci/l, and even after 5 years the concentration would still be above the maximum acceptable level, with a value of 1.9×10^{-7} Ci/l (taking dispersal into account).

8.3. Calculation of dose loads

It is possible to estimate the effects of the impact of radioactive contamination on water body ecosystems more accurately by first analyzing the dose loads of ionizing radiation received by hydrobionts. In our calculations we will take into account the contribution of two components: external radiation from radionuclides distributed throughout the water, and irradiation as a result of the decay of incorporated radionuclides. The method of calculation is contained in the appendix, with the concentrations taken from Table 1.

Dose loads were calculated for organisms belonging to trophic levels of phytoplankton and zooplankton (Table 3), since the upper layers of the water body would be most severely contaminated at the initial moment. The further redistribution of radionuclides throughout the volume of the water body and their rapid decay would lead to lower dose loads for organisms belonging to the remaining trophic levels. In addition, trophic levels of phyto- and zooplankton form the foundation of the ecological pyramids of water bodies and any impact on them would have a significant influence on the ecosystem as a whole. It should be noted that even organisms belonging to the same trophic level would strongly differ in their size and capacity for accumulating radionuclides. Hence Table 3 shows intervals, with the differences in accumulation coefficients and size of organisms being taken into account within their limits. The levels of dose loads received by hydrobionts in water bodies located in zones of severe or moderate contamination contained in Table 3 allow us to conclude that radioactive contamination would have a substantial impact on these ecosystems, since the levels of irradiation of hydrobionts in these cases would lead to dose loads near LD_{50} (Table 4). In addition it should be emphasized that we considered a case where radioactive con-

Table 3. Dose loads received by hydrobionts over a period of 1 day (Gy)

	Zone of intensive contamination			Zone of moderate contamination			Zone of weak contamination		
	1 day	10 days	1 year	1 day	10 days	1 year	1 day	10 days	1 year
Phyto-plankton	1–50	0.1–5	$(0.1–5)\cdot 10^{-2}$	0.2–10	$(0.2–10)\cdot 10^{-1}$	$(0.2–10)\cdot 10^{-3}$	$(0.7–35)\cdot 10^{-1}$	$(0.7–35)\cdot 10^{-2}$	$(0.7–35)\cdot 10^{-4}$
Zoo-plankton	1–100	$(0.5–50)\cdot 10^{-1}$	$(0.1–10)\cdot 10^{-2}$	0.2–20	$(0.1–10)\cdot 10^{-1}$	$(0.2–20)\cdot 10^{-3}$	$(0.7–70)\cdot 10^{-1}$	$(0.3–30)\cdot 10^{-2}$	$(0.7–70)\cdot 10^{-4}$

Table 4. Levels of lethal doses from a single irradiation for hydrobionts belonging to different phylogenetic groups [40]

Group of organisms	Dose (Gy)	Type of estimation
Bacteria	45–7550	LD_{90}
Blue-green	4000–12000	LD_{90}
Other species of algae	30–1200	LD_{50}
Protozoa	6000	$LD_{50/30}$
Mollusks	200–1050	$LD_{50/30}$
Benthos	15–366	$LD_{50/30}$
Fish	11–56	$LD_{50/30}$

1 Gy = 100 rad

tamination was uniformly distributed with the chosen territories (severe, moderate, and weak contamination). Actually, since a nuclear exchange would occur mainly in industrial centers located near the sources of water supply, the levels of contamination of water bodies used for the water supply would obviously be substantially higher.

In addition it should be noted that dose loads are calculated without taking into account radioactive contamination as a result of the destruction of nuclear power plants.

The additional contamination of water bodies by products of nuclear reactors and radioactive waste would lead to a significant increase in dose loads for hydrobionts, which would exceed the corresponding LD_{50} for many species. This would lead to the complete or partial degradation of the ecosystems, retarding the self-cleaning in water bodies, and a deterioration of the water quality.

8.4. Contamination of freshwater body systems by heavy metals

Heavy metals emitted into the atmosphere as a result of "fires of civilization" would gradually settle to the earth. Part would be absorbed by the soil, and part would be dissolved in water and would enter freshwater reservoirs.

Here we will use the same hypotheses used earlier for estimations of loads in a standard water body system. Thus, in addition to direct settling on the surface of the water body, approximately 10 % of the total mass falling on the water shed would be washed into the water body by surface flow. In contrast to radioactive contamination, heavy metal salts do not decay, remaining intact for any amount of time, and, when leached out, may gradually find their way into the water body. Nevertheless, on account of filtration and ion exchange, they may be retained by the soil and underlying surface to a significant degree. If we use, by analogy, data for salts of other metals (Al, Fe, etc.), we find the range of values to be quite great [35]: from 0.4 to 15 % are retained by the soil. Also by analogy we recall that these values for cesium and strontium are 99.5 and 98.5 %, respectively [78]. All this significantly complicates the estimation of the release of heavy metals along with groundwater into the water body. Obviously their release would be more substantial in the initial period when concentrations of heavy metals in filtered salts would remain high. Thereafter the amount cor tributed by ground-water to the load entering the water body would fall rapidly (exponentially). Thu we can assume that in the first 20 days an additional 10 % of the fallen mass of conservative pollutant would be carried into the water body by ground-water. Thereafter, the concentrations of heavy metals in surface flow can be considered to remain at a level not exceeding the concentratior in the water body.

Since the characteristic renewal time of the mass of water in a lake (reservoir) is substantially higher than this period and than the characteristic time it takes for contamination to fall out (about 7 hours)*, we can assume that all contamination accumulated in this way would remain in the lake for a rather long period of time. The level of contamination can be calculated using the following formula:

$$C = (9k + 1)\, p/1, \qquad (1)$$

where the first term in parentheses corresponds to the collection from the watershed (k = 0.2 – percentage of washed off contamination), the second one to the settling onto the water body itself, and p is equal to the level of contamination falling on the region of interest ($p = p_1 n$, where n is the population density, p_1 is the contamination on territories with a density of n = 1 ind/km^2). The results of these calculations are contained in Table 5.

8.5. Conclusion

It has been shown that land supplies of fresh water would be unfavorable for human use as a result of radioactive contamination for an extented period of time after a nuclear conflict (5 years or more). In addition, water body ecosystems would suffer significantly, with their capacity for

* Renewal time can be calculated as $R = 1/m \cdot r(yr)$, where l = depth, r = average annual precipitation, m = ratio between watershed area and water body area. In the given case l = 4 m, r = 30 cm/yr, m = 10 and R = 1.3 yr

Table 5. Content of heavy metals in freshwater bodies after fires

1. Element	Hg	Pb	Cd	As	Se	Cu	Zn
2. Emissions (ρ_1) (mmg/m^2) for population density 1 ind./km^2	6	500	40	4000	0.1	700	4000
3. Emissions (ρ) (mg/m^2) for population density 200 ind./km^2	1.2	100	8	800	0.02	140	800
4. Top acceptable levels in potable water (mg/l)	0.005	0.1	0.6	0.05	–	1	18 1.1 (for fish)
5. Concentration in lakes and reservoirs (mg/l) for population density 1 ind./km^2	$3.6 \cdot 10^{-6}$	$3 \cdot 10^{-4}$	$2.4 \cdot 10^{-5}$	$2.4 \cdot 10^{-3}$	$6 \cdot 10^{-8}$	$4.2 \cdot 10^{-4}$	$2.4 \cdot 10^{-3}$
6. The same for population density 200 ind./km^2	$7.2 \cdot 10^{-4}$	0.06	$5 \cdot 10^{-3}$	0.5	$1.2 \cdot 10^{-5}$	$8.4 \cdot 10^{-2}$	0.5

natural self-cleaning undoubtedly being affected. The process of renewal of biohydrochemical cycles would take even longer. It is difficult to predict the new homeostatic state of freshwater ecosystems. Contamination by heavy metals would be retained for a relatively long period of time, but would be rather low: the maximum acceptable level would be exceeded only for arsenic if the population density were not less than 10 ind./km^2 (possibly for tin as well).

In addition to "contaminating" effects, the freshwater cycle would also be damaged by "destructive" impact. This is connected first of all with the unavoidable destruction of hydroelectric dams. As important economic objects, these would be independent targets of nuclear attack; on the other hand, they are generally located near densely-populated cities. The destruction of reservoirs means the elimination of huge supplies of fresh water. In addition, the simultaneous overflowing of these supplies would cause flooding by contaminated water.

Finally, freshwater ecosystems can undergo substantial changes as a result of so-called "nuclear winter". The freezing of water bodies may come about as a result of prolonged frost. The determination of the ice thickness on water bodies and streams can be reduced to the solution of a rather complex problem for a system of partial differential equations with a movable boundary. We can fall back on empirical formulas which nevertheless must take into account a large number of factors: temperature, air humidity and wind velocity on the surface of the ice, thickness and density of snow, direct and diffuse radiation, and water velocity and temperature below the ice. For our approximate estimation we use Byndin's simple formula [13a].

$E = \varphi\sqrt{\Sigma\theta}$, where $\varphi = 2–11$, usually $\varphi = 3.4$, $\Sigma\theta$ is the sum of negative temperatures. According to this formula the ice thickness would be on the order of 3.5 m for an extended "nuclear winter" of 250 days with an average temperature of –25 °C, which would lead to the complete freezing of many water bodies and streams.

8.6. Appendix

8.6.1. β-radiation

Standard equations from [9] and [58] were used in the calculations. The amount of the absorbed dose in a homogeneous medium with uniformly distributed β-sources with activity q (Ci/kg), average energy

$\bar{E}_\beta$ (MeV) located farther than $R(E_\beta)$ from the boundary ($R(E_\beta)$ — actual path length for β-particles with boundary energy E_β), may be calculated with the formula:

$$p_1^\beta = 0.591\, qn\bar{E}_\beta \frac{\text{rad}}{\text{s}}, \quad (2)$$

where n = yield of β-particles per fission.

To calculate the dose from β-radiation of incorporated radionuclides the Levinger equation is usually used. It characterizes the doese yield from an isotropic point-source of β-particles in boundless medium:

$$D(r) = \frac{K}{(\nu r)^2} \{C[1 - \frac{\nu r}{c} \exp(1 - \frac{\nu r}{c})] + \nu r \exp(1 - \nu r)\},$$
$$[1 - \frac{\nu r}{c} \exp(1 - \frac{\nu r}{c})] \equiv 0 \quad \text{for } \nu r \geqslant c. \quad (3)$$

Here D(r) is the dose yield produced by a source of β-particles at the distance r(cm), ν (g/cm^3) — is the effective absorption parameter for β-particles, c — unitless parameter, depending on the energy of β-particles, K — norming term determined by the condition that the total energy absorbed in infinite volume be equal to the average energy of β-radiation per fission. As shown in [89], these parameters take the following values: $K = 1.28 \times 10^{-9} \rho^2 \nu^3 \bar{E}_\beta \alpha$ rad/fission, $\alpha = [3c^2 - (c^2 - 1)\, e]^{-1}$,

$$\nu = \begin{cases} 15\, E_\beta^{-1.7} (2 - \frac{\bar{E}_\beta}{\bar{E}_\beta^*}), & \bar{E}_\beta \leqslant 0.5 \text{ MeV}, \\ \frac{18.6}{(E_\beta - 0.036)^{1.31}} (2 - \frac{\bar{E}_\beta}{\bar{E}_\beta^*}), & \bar{E}_\beta > 0.5 \text{ MeV} \end{cases}$$

$C = \{3.11,\ E_\beta \leqslant 0.1$ MeV; $1.35\, E^{-0.364},\ 0.1 < E_\beta \leqslant 2.25$ MeV; $1.0,\ E_\beta \geqslant 2.25$ MeV$\}$,

where E_β = boundary energy of the β-spectrum (MeV), $\bar{E}_\beta^*$ = mean energy of the hypothetically allowed spectrum with boundary energy E_β per fission. For the allowed β-spectrum $\bar{E}_\beta / \bar{E}_\beta^* = 1$.

The distribution of the dose yield interior and exterior to sources of various forms results from integration of (3) in the appropriate coordinates. Thus, the dose yield at the center of a spherical source of β-particles with activity uniformly distributed over the volume is specified by equation [58]:

$$p_2^\beta = 0.591\, q^* n\bar{E}_\beta \{c^2\alpha[\frac{\nu R}{c} + (1 + \frac{\nu R}{c}) \exp(1 - \frac{\nu R}{c}) - 3] +$$
$$+ 1 - \alpha(1 + \nu R) \exp(1 - \nu R)\} \frac{\text{rad}}{\text{c}} \quad (4)$$

where q^* = concentration of radionuclides in the source material (Ci/kg); R = radius of the source (cm).

8.6.2. γ-radiation

The standard expression for the dose yield inside an infinite source of γ-quanta with uniform activity [48] may be used to calculate the dose yield from external irradiation of hydrobionts in water:

$$p_1^\gamma = \frac{4\pi\Gamma\gamma_2}{\mu_B \gamma_1} q_\nu [\frac{A}{1+\alpha_1} + \frac{1-A}{1+\alpha_2}], \quad (5)$$

where μ_B, γ_1 = attenuation and absorption parameters for γ-quanta in water, respectively: γ_2 = absorption parameter for γ-quanta in the air; q_ν = radionuclide concentration in water; A, α_3, α_2 = Taylor expansion coefficients for the accumulation factor; Γ = gamma-constant.

Since usually path lengths of γ-quanta in biological tissue by far exceed the sizes of hydrobionts, the contribution of irradiation by incorporated radionuclides was neglected in the estimation of the total irradiation dose yield of hydrobionts.

8.6.3. Numerical values of parameters

In the calculations we took into account changes in the isotope composition. For radionuclides entering water at each rated moment in time, calculations were made for the mean energy of the β-spectrum and of γ-quanta per fission.

Numerical values for parameters were taken from: [38] – for β-radiation (the β-spectra were considered to be allowed), [48] – for γ-radiation.

Parameters for radionuclide accumulation, which we take as the ratio between radionuclide concentrations in organisms of hydrobionts and in water, are chosen from data presented in [70], [47], and [42]. The element composition of the fission product mixture was taken into account. Accumulation parameters for radionuclides of the same chemical group were considered to be the same and equal to the mean coefficient with respect to all radionuclides of this group.

9. NUCLEAR CONFLICT, AGRO–ECOSYSTEMS AND THE FOOD SUPPLY PROBLEM

9.1. Introduction

In examining the impact of a nuclear war on agro-ecosystems, it is necessary to add factors such as the destruction of the industrial-energy potential of developed countries to the factors having a direct influence on productivity, which we have already considered. "World economy" and "world trade and exchange" would cease to exist; furthermore [104], in developed countries taking part in the conflict the restoration of the potential to its previous level would be impossible. All this would have a direct effect on the functioning of world agriculture.

9.2. Consequences for the agro-ecosystems of developed countries

Consequences would be most severe for the agro-ecosystems of developed countries.

Highly productive agriculture in these countries is supported by a whole series of factors:

- wide use of agrotechnology, agricultural machinery and structures;
- use of modern agrotechnology (mainly use of fertilizers, herbicides and insecticides). The annual use of fertilizers (NPK) in developed countries is close to 80 million mt, more than 120 kg/ha (data for the early 1980's) [92];
- measures for improving agro-ecological potential (melioration, irrigation, etc.).

In developed countries, from 1.5 to 3.5 units of "artificial energy" [87] (fuel, machinery, fertilizers, pesticides, etc.) are expended in the production of one unit of agricultural product. The corresponding percentages of expenditure of "artificial energy" for the production of corn in the USA is as follows [115]: fertilizer – 39 %; fuel 19 %; machinery – 14 %; drying – 10 %; irrigation 9 %; herbicides – 2 %; others – 7 %. Human labour comprises only 0.1 %, which allows the percentage of the economically active population employed in agriculture in developed countries to be reduced to a few percent (in the USA – 2 %; average for developed countries – 11.6 % [76]).

The destruction of the economic potential of developed countries would lead to an end of the flow of "artificial energy" into agriculture and a transition to primitive forms characterized by low productivity and a high demand for labour force. A redistribution of labour resources into agriculture would take place, which would retard the restoration of the economy. While developed countries are currently the major producers and exporters of agricultural products (62 % of world agricultural products, 70 % of world wheat exports, etc. [76]), after a nuclear conflict they would only be able to feed their own surviving population.

9.3. The food supply problem in developing countries

Let us consider the consequences of a nuclear conflict on agriculture and the food supply in developing countries, which is related to the cessation of trade, economic aid from developed countries and the destruction of the system of production redistribution.

Agricultural production in developing countries is characterized by a dependence on the economies of developed countries. Agricultural technology and machinery is almost completely imported (import of tractors in 1981 was approximately 170 thousand [94]). Fertilizers in developing countries are in short supply (averaging 10–15 kg/ha), with a significant percentage being imported from developing countries (in 1982 imports of nitrogen fertilizers comprised 26 % of the total amount of nitrogen fertilizers used; phosphorous – 30 %; potassium – 98 % [92]). Imports account for 100 % of chemical means of plant protection (pesticides, herbicides, etc.) used.

Agriculture in developing countries remains traditional, retaining primitive agrotechnology and characterized by low productivity. The production by one agricultural worker in developing countries is 10 times less than in developed countries. Despite of the fact that on the average approximately 60 % of the economically active population is employed in agriculture in developing countries, supplying the population with food is a major problem. According to FAO data, the level of nutrition of 435 million people in developing countries is below the minimum level necessary for the maintenance of life, established by the World Health Organization and the FAO [76] (to allow for normal functioning and growth of children and adolescents and for light activity of adults, FAO and WHO have determined this to be 1.5 BMR (Basal Metabolic Rate – energy needed for the maintenance of body weight under the condition of resting, or about 1900 kcal for the small adult male) [76] (in 1980 the population of developing countries was 3252 million people or 74 % of the world population). The average nutrition level in developing countries was 2300 kcal per person per day, with vegetable foods making up more than 90 % (in developed countries the corresponding values are 3400 kcal and 69 % [93]).

A significant portion of food products are imported from developed countries. Grain imports from developed countries at the end of the 1970's and beginning of the 1980's amounted to approximately 80 million tons; 35 million tons were imports of wheat alone (or 40 % of wheat production in developing countries [76]). According to FAO predictions, the period from now until the year 2000 will be a period of intensification of agricultural production in developing countries [76]). The flow of "artificial energy" directed toward agriculture will grow 5 times during this period, with more than 90 % going toward fertilizers and agrotechnology (Figure 22).

The use of fertilizers will be increased 7 times; the tractor fleet will be increased 8 times (see Table 1) [76]. On account of this intensification the grain yield can be expected to rise more than 1.5 times (wheat yield will increase from 1.29 t/ha to 2.20 t/ha; rice yield will increase from 1.95 t/ha to 3.1 t/ha).

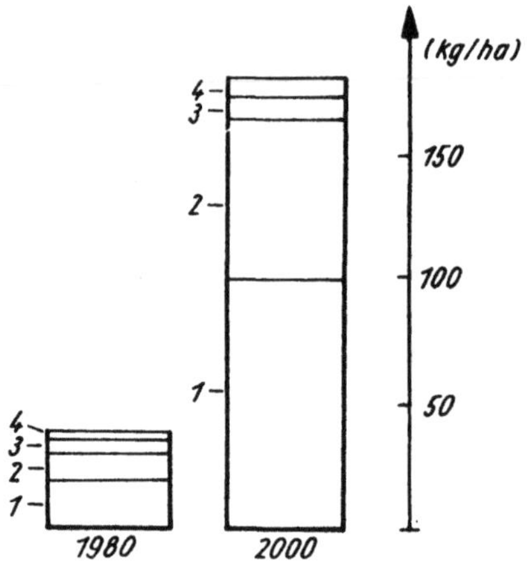

Fig. 22. Flux of "artificial energy", used by plant-growing industry in developing countries (in oil equivalence), supposing the current tendencies are kept
1 – fertilizers; 2 – agrotechnics; 3 – reclamation; 4 – pesticides

Table 1. Intensification of agricultural production in developing countries

	1974–75	2000
Fertilizer (thousand tons NPK)	13.485	90.000
Tractor fleet (thousand tractors)	1.908	12.000

The process of intensification will be sustained by active economic support from developed countries.

However, in spite of the measures being taken, the growth rate of agricultural production in developing countries will lag behind the growth in demand. According to FAO predictions, if current development trends continue, the grain deficit in developing countries will be increased, and gross imports in the year 2000 will be 226 million tons [76]. While there will be an absolute growth in the production of agricultural products, indicators such as self-sufficiency of developing countries would fall (data for grain is contained in Table 2).

The elimination of industrial and energy potential in developed countries as a result of a nuclear conflict would lead to the destruction of economic relations with developing countries, and the cutting off of exports of agrotechnology, fertilizers, pesticides, herbicides and food products. Table 3 contains an estimation of grain production in developing countries after a conflict, taking into account the fact that the cutting off of imports of agrotechnology and fertilizers would lead to an end in the growth of yield, while the absence of pesticides and herbicides would lead to a loss of 30 % of the harvest from pests and disease (losses in harvest for various cultures are contained in Table 4).

Table 2. Grain production in developing countries under current development trends (in millions of tons)

Average 1975–79	total	Africa	Far East	Latin America	Near East
Production	382	43	207	80	52
Demand	415	51	219	81	64
Self-sufficiency (%)	92	83	95	98	82
Calorie supplies (kcal/capita/day)	2179	2180	2023	2525	2562
Year 2000	**total**	**Africa**	**Far East**	**Latin America**	**Near East**
Production	636	61	358	146	71
Demand	768	110	384	168	106
Self-sufficiency (%)	83	56	93	87	67
Calorie supplies (kcal/capita/day)	2363	2306	2208	2698	2848

Table 3. Annual grain production in developing countries after a nuclear conflict before the year 2000 (10^6 tons)

Total	Africa	Far East	Latin America	Near East
267	30	145	56	36

Table 4. Loss in harvest from pests and disease [50]

	Collected harvest (mil. t./year)	Losses (mil. t./year)
Grains (wheat, rice, buckwheat, millet, corn, barley, rye)	960–1000	500–510
Sugar beets	210–250	65–75
Sugar cane	470–480	550–570
Potatoes	270–280	125–735
Fruit	66–67	21–22
Vegetables	202	78–79

If current demographic trends in developing countries continue (Table 5), the disproportion between production and demand for agricultural products will grow rapidly. Table 6 contains estimations of self-sufficiency in agricultural products for developing countries. Table 7 contains an estimate of the population potentially possible in developing countries after a conflict (i.e., the population of these countries will be in a position to supply themselves with food products at a level of 1900 kcal/day, i.e. the level of basic metabolism) [76].

This was calculated under the assumption that a reduction in grain production would lead to a proportional decrease in the calorie supply for a constant population, or to the proportional reduction in the population for a constant calorie supply.

Table 5. World population 1980–2000

Population (millions)	1980	1990	2000
World	4,415	5,275	6,199
Developing countries	2,259	2,906	3.630
Developing countries (incl. China)	3,252	4,027	4,87
By region			
Africa	370	503	668
Far East	1,313	1,653	2,068
Latin America	363	472	601
Near East	213	278	353

Table 6. Food self-sufficiency (%) in developing countries in the year 2000 under continuation of current demographic trends

Calorie supplies	total	Africa	Far East	Latin America	Near East
Basic metabolic rate (1900 kcal/capita/day)	44	33	44	47	51
level in 1980's	38	29	41	36	38
level in 2000	35	27	37	33	34

Table 7. Estimation of potentially possible population in developing countries after a conflict (calorie supply of 1900 kcal/day)

Population (mil. ind.)	Developed countries, excl. China	Africa	Far East	Latin America	Near East	China
	1679	250	926	333	162	870
% of 1980 population	74	67	70	91	75	85

9.4. Conclusion

a) In developed countries there would occur a degradation of current agro-ecosystems, their complete destruction and a transition to primitive forms of agriculture, characterized by extremely low productivity. This situation would remain for an extended period of time, since the redistribution of labour resources from industry to agriculture would greatly retard the process of restoration of the industrial-energy potential in these countries.

b) The cutting off of imports of agrotechnology, fertilizers, herbicides, and pesticides from developed countries by developing countries after a conflict would lead not only to an end in the growth of agricultural production, but also to an absolute decrease in production in developing countries. By the year 2000, given the continuation of current demographic trends, the food self-sufficiency of developing countries would be about 40 %. Thus a nuclear conflict between developed countries would lead to a global food crisis.

10. DEMOGRAPHIC CATASTROPHE

10.1. Introduction

In the present chapter we will be concerned neither with the thermic effect nor with the striking action of the blast wave, both of which would exert a lethal impact on nuclear weapon target areas (after a nuclear attack of a 5,000–10,000 Mt yield about 750 million to 1.1 billion people would be immediately killed, mainly in the northern hemisphere [90]). The consequences of climatic changes would be no less grave: a sharp cold snap, decrease in light intensity, extinction of many plant and animal species, starvation due to the annihilation of food supplies, lack of crops for a period of at least one year, erosion and contamination of agricultural areas, spreading of diseases, and other stresses would cause the death of about 2 billion people in the northern hemisphere, where, perhaps, no one would survive. The possibility of the complete extinction of Homo sapiens (due to the above-mentioned factors) should not be excluded.

Here we will be more interested in the long-term consequences of the impact of a nuclear war on the human population, as well as in the possibility of their surviving under altered conditions of the social and natural environment. The biological consequences of a nuclear war for the human population can be divided into two main categories: the *medical* (or *somatic*) and *genetic*. We will consider each of them separately.

10.2. Somatic (medical) consequences of a nuclear conflict for the human population

The medical, or somatic, category can, by its nature, be related to health problems (and, as a result, to an increased level of mortality, decrease in fertility, and birth of defective progeny). The somatic consequences would be most apparent immediately after the conflict. This category of damage would be the most serious for people, ranging from lethal and sublethal levels immediately after a conflict to the more remote influences on the health of those who would survive the conflict (increase in the risk of future malignant neoplasms, weakening of resistance to diseases, damage to various vital systems of an organism, negative influence on progeny).

One should not think, however, that the somatic category of damage would be restricted to the life-span of a single generation. Even if we ignore the radioactive contamination of the environment (which would have a prolonged impact on the population), the deterioration of natural conditions, etc., the weakening of health alone would be enough to bring about unfavourable effects on the normal course of pregnancy, the birth of normal progeny, and eventually on children's health (possibly for a number of generations). Nevertheless, typical of the somatic category would be a noticeable decline in the severity of the consequences of a nuclear war over a number of generations, provided the environmental stress were removed. This distinguishes somatic damage from genetic damage (considered below), which can be transmitted from generation to generation far longer, in many cases having no explicit effect on the carriers of genetic anomalies, but harming their progeny.

A great amount of toxic substances would enter the atmosphere as a result of nuclear explosions and the fires accompanying them. Calculations show that a tenfold increase in atmospheric contamination by combustion products can bring about a catastrophic growth of allergic diseases, and diseases of the nervous system and organs of sense and respiration. Practically every person would become chronically ill [7].

Lack of medical care, severe environmental conditions (e.g. cold, decrease in illumination, starvation), together with the disruption of social institutions would make the survival of chronically ill or weakened persons, as well as healthy persons, questionable. It is well-known that severe living conditions themselves increase the likelihood of illness. For example, the siege of Leningrad during World War II had a severe effect on the health of its inhabitants, especially women and children. In 1942 the rate of illness of the labour population increased by 25 % as compared with 1940; the average duration of disease, an indication of its severity, increased two times. Women's health problems manifested themselves in the form of amenorrhea (64–95 %), and in pre- and postnatal illness with a high degree of mortality. Almost all children showed signs of alimentary dystrophy, metabolic disturbances, or severe functional disorders of the nervous and cardiovascular systems, intestinal tract or kidneys (cit. [7]). The influence of toxic contamination of aquifers and thus of drinking water as a result of nuclear explosions should also be noted here.

A partial destruction of the ozone layer would cause an increase in UV-radiation after the restoration of atmospheric transparency. The irradiation would repress the immune systems of mammals (creating favourable conditions for the spread of diseases) and would bring about the spreading of blindness in man and other mammals.

Ionizing radiation, both direct radiation (from nuclear explosions) and radiation carried by air currents and thus having a global character, also belongs to the somatic category. After a nuclear exchange with a yield of 5742 Mt, irradiation within an area of 5 mill. km^2, over a 48 hour period, would be on the order of 1000 R and above, which would be sufficient to bring about the death of all irradiated human beings. Irradiation of 30 % of the land surface in the middle latitudes by 500 R or more γ-radiation would cause a sharp increase in mortality in all mammals, including man [90].

The maximum dose survivable by man is on the order of 600–700 rad [56], a dose of 500–600 rad causes permanent male sterility; permanent infertility can be caused by a dose of 400 rad for young women and by a dose of 250 rads for women over 35 [68].

According to other data the lethal dose of radiation for a 48 hour period is 350–400 R [90]. An acute dose of γ-radiation would cause a mortality rate of 50 % of the adult population during a 60-day period. Ionizing radiation has a teratogenic effect on embriogenesis; it also increases the risk of leukemia and tumors as a result of fetal irradiation.

A γ-radiation dose on the order of 150 rad would most likely not be lethal, at least among the physically healthy portion of the population. However, approximately 50 % of irradiated humans would show signs of gastro-intestinal disorders. A dose of more than 200 rad would disturb hemopoiesis [84].

Irradiation in doses of more than 200 rad would have a detrimental effect on vision, with cataracts appearing in the course of time. Irradiation stimulates the appearance of malignant neoplasms [84].

According to the data presented in Chapter 1, the accumulated dose over 24 hours would exceed the lethal dose in regions with a population density of 150 or more ind./km^2, and would cause infertility in women in regions with a density of 30 or more ind./km^2. In regions with a population density of about 10 ind./km^2, the accumulated dose would most likely not be lethal, but irradiation would cause some functional disorders. The above consequences correspond to the influence of acute irradiation during a 24-hour period and would be aggravated by the further influence of ionizing radiation and "ordinary" pollution.

10.3. Human population dynamics after a nuclear war

Let us evaluate the demographic dynamics of the remaining population under changed conditions of the social and natural environment. In these calculations we will hypothesize that the destruction of systems of social welfare and medical care would result in demographic properties of the human population (fertility and mortality) on at least a medieval level. However, the additional influence of the negative effects of ionizing radiation would bring about a permanent decline in the human population and its eventual extinction.

We can estimate the rate of variation in population using a model of the population according to age (three age groups of 20 years each). As a first approximation we will take into account only the decrease in fertility f due to ionizing radiation. Since there are no exact quantitative data on the effects of radiation on fertility, we will assume that persons receiving a dose D of 600 rad/yr would be rendered completely sterile. We will also assume a linear dependence of fertility on dose. Using these assumptions we can find the growth rate of the population. For doses where the calculated rate is less than 1, a decrease in population and the eventual extinction of the population can be expected.

Thus, we assume that "fertility" depends linearly on the average dose of radiation for one year $f = K \cdot D + \alpha$, where f = fertility, and D = radiation dose. The parameters k and α are determined by the assumption that doses of 600 rad/yr would bring about sterility. Thus

$$f(D) = \frac{-f^*}{600} D + f^*, \qquad (1)$$

where f^* is the birth rate in the absence of irradiation.

To determine the rate of population growth we can use a simple model of the type used by Leslie. The human population can be divided into 4 age groups: 0–20 years, 21–40 years, 41–60 years, and over 60. The Leslie operator can be written approximately as follows:

$$\begin{matrix} 0 & f^* & 0 & 0 \\ p_1 & 0 & 0 & 0 \\ 0 & p_2 & 0 & 0 \\ 0 & 0 & p_3 & 0 \end{matrix} \qquad (2)$$

where f^* = birth rate of the second age group, and p_i ($i = 1, 2, 3$) is the probability of transition from a group i into the group (i + 1) (survival probability of the corresponding group). The eigenvalues for (2) can be determined from the equation $\lambda^2(\lambda^2 - p_1 f^*) = 0$. The maximum eigenvalue is

$$\lambda_{max} = \sqrt{p_i} \cdot f^* \qquad (3)$$

We estimate the probability of living till the age of 20 using data contained in [99].

By averaging the available data we get $p_1 \approx 0.58$; $p_2 \approx 0.62$; $p_3 \approx 0.38$.

Table 1. Survivors per 100 newborns (all mortality factors)

	Age	1330–1479	1480–1679
a) men	0	100	100
	20	54	54
	40	28	33
	60	8	15
b) women	0	100	100
	20	61	65
	40	43	44
	60	19	17

According to (3), we need to know f^* in order to calculate the population growth rate λ_{max}. As we did not succeed in finding reliable data on fertility in the Middle Ages, we use the following approach to estimate f^*. We will use data from the book [103] on the human population dynamics in Europe from 1100 to 1600. The growth rate is calculated using the ratio of population sizes in two successive periods of time. Corrections are then made so that the rate corresponds to a unit of time of 20 years (used in the Leslie model). Finally, the average rate is found (using the corresponding time periods as weight coefficients for the rates of individual periods):

Table 2. Population dynamics and growth rate in Europe from 1100 to 1600

Year	Population (mill.)	$\lambda_i = e^{\frac{20}{\Delta T} \ln \frac{N_{i+1}}{N_i}}$	mean weighted value λ
1100	35		
		$1.06 = \lambda_1$	
1200	49		
		$1.06 = \lambda_2$	
1250	57		
		$1.08 = \lambda_3$	
1300	70		
		$1.02 = \lambda_4$	$\lambda = 1.03$
1340	74		
		$0.88 = \lambda_5$	
1400	52		
		$1.05 = \lambda_6$	
1500	67		
		$1.05 = \lambda_7$	
1600	89		

Thus, according to the scheme in Table 2 the average growth rate λ is equal to 1.03. Knowing λ we can find f^*, which will be equal to 1.82.

Using (1) and (3) we can find the dependence of the growth rate on the dose of irradiation according to our assumptions as follows:

$$\lambda(D) = \sqrt{p_i}\,(-\frac{f^*}{600}D + f^*) \qquad (4)$$

Formula (4) allows us to find the conditions under which the maximum eigenvalue passes through 1. If we let $p_i \approx 0.58$, the critical radiation dose D = 32 rad/yr (the population declines beginning with this dose). If we assume that the dose causing sterility is equal to, say, 1200 rad/yr, rather than 600 rad/yr, we can easily find that given these demographic parameters the critical dose would be $D \approx 60$ rad/yr (Fig. 23).

Let us now assume that "fertility" is constant for a value of D from 0 to 100 rad/yr, and, beginning with 100 rad/yr, decreases linearly for growth in D; then

$$f(D) = -\frac{f^*}{500}D + \frac{6}{5}f^*.$$

Given this assumption, if we consider doses causing sterilization to be 600 rad/yr and 1200 rad/yr, we get critical dose values of 145 rad/yr and 290 rad/yr, respectively.

Since the minimum radiation dose per year (for 2, 3, etc., years after a nuclear war) for territories in the USA, Canada, Europe (excl. USSR) and the European part of the USSR (see Chapter 1) corresponds to an average population density of 10 ind./km² (i.e., the vast majority of the population would receive substantially larger doses) and varies from 200 to 1000 rad/yr, it can be concluded that even in areas with a relatively low population, which would be least subject to nuclear attacks, the demographic dynamics would be negative – the human population would die out within several generations.

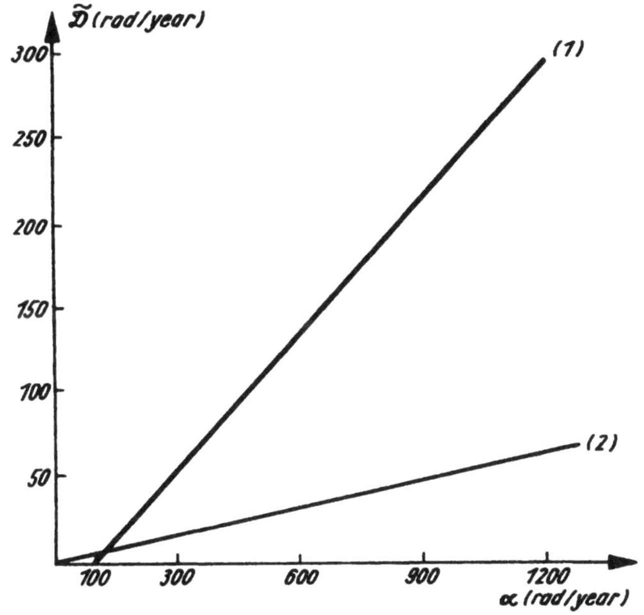

Fig. 23. Graph showing the dependence of the "critical" dose of irradiation (D) on a value of sterilizing dose (α). The "critical" dose is a dose above which a decrease in population growth would begin
(1) – Under the assumption that "fertility" would not change for values of D up to 100 rad/year;
(2) – under the assumption that "fertility" decreases starting with a zero dose

Other factors would aggravate the situation: hunger, a high level of "ordinary" pollution, poisoning of drinking water supplies, etc. It should be noted that our hypothesis on the medieval pattern of mortality is rather "optimistic". Also, medievel man had adapted to his conditions of existence, while modern man would be poorly adapted to medieval conditions.

10.4. Outbreak of epidemics

Using rather simple mathematical models we will try to estimate the probability of an outbreak of epidemics of the most dangerous diseases for the population surviving a nuclear war.

We will use the AMBIO data as a basis for determining the percentage of survivors after nuclear attacks [78]; according to this data, nuclear attacks would occur mainly in cities with a population of not less than 100 thousand inhabitants. As can be seen from these data (see Figure 24), the survival ratio would be approximately the same in all cities regardless of size, and would be equal to $\delta = 0.2$. If we consider that the urban population lives mainly in cities of not less than 100 thousand, and if we assume that the percentage of rural residents to die in a war would be extremely low, the number of survivors in any country can be estimated using the following formula

$$N_* = N_0 \, (1-(1-\delta)k) \qquad (5)$$

where N_0 is the population in the country before the beginning of the war, and k is the percentage of urban population. Thus, for example, for Great Britain, according to 1982 data, $N_0 = 54.2$ million people, k = 90.8 %, whereby $N_* = 14.8$ million, or 27 % of the initial population.

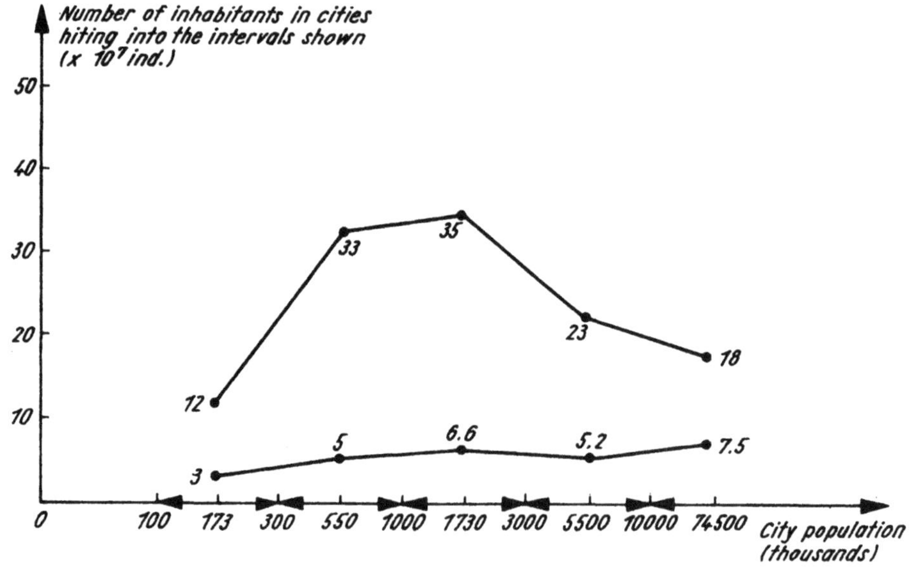

Fig. 24. Distribution of population among cities with various numbers of inhabitants in regions of the conflict. Data were obtained on the basis of estimation of the number of cities in certain intervals (see AMBIO scenario) and the average population for an interval (see the upper graph). The lower graph shows the number of survivors after a war. The sum of the values on the lower graph is equal to 0.2 of that of the upper graph

In terms of applying this to problems of outbreaks of epidemics, it is worthwhile to look at the question of decline in population density after a nuclear exchange. In addition to the above, we will use several more hypotheses, the first relating to the redistribution of urban population. It is assumed that because of the impossibility of survival under urban conditions, the population would spread over rural territories. In addition, it is assumed that these migrations would cover relatively small distances ($10^1 - 10^2$ km). Therefore, we can say that soon after a nuclear attack the population would be distributed relatively uniformly in accordance with the estimations given below. Finally, the most restrictive hypothesis is that the population density after a nuclear exchange would be determined solely by population density beforehand, i.e., attacks would be distributed over the territory in such a way that the survival rates in areas of equal population density would be equal. This hypothesis is supported by data on the relationship between the mean population density and the percentage of urban population in a number of European countries in 1982 (see Table 3), as well as by the AMBIO scenario, which shows a uniform distribution of attacks in cities of equal population.

Although the data contained in Table 3 do not allow us to assign any simple dependence to the function k(n), where n is the population density, it can nevertheless be seen that for increasing n-values there is a saturation level for k corresponding to an asymptotic value of $k_A = 0.9$. As n decreases, k also decreases though not falling below $k_0 = 0.4$. This enables us to approximate the data contained in Table 3 by means of a picewise linear function

$$k(n) = \min (k_0 + rn, k_a). \quad (6)$$

The calculations show that r should be assigned a value of 0.003 $(\text{ind./km}^2)^{-1}$. The corresponding graph is given in Figure 25. It can be seen that points falling outside the curve correspond to countries with extremely non-uniform population density, i.e., countries which are not representative under the given assumption.

Table 3. The relationship between percentage of urban population and mean population density

Country	Density (ind./km^2)	Percentage urban population
Austria	90	54.1
Albania	100	36.8
Belgium	324	72.4
Bulgaria	80	62.5
Great Britain	222	90.8
Hungary	115	53.7
German Democratic Republic	155	76.3
Denmark	119	84.2
Italy	82.4	69.3
the Netherlands	347	88.1
Norway	12.7	43.7
Poland	116.6	58.7
Rumania	95	49.6
Federal Republic of Germany	241	84.7
Finland	14.2	62.2
France	98.5	77.9
Czechoslovakia	61.7	69.2
Switzerland	157.4	58.1
Sweden	19	87.2
Yugoslavia	88	47.3
Japan	320	76.2
Canada	24.8	80.1
USA	25	72.9

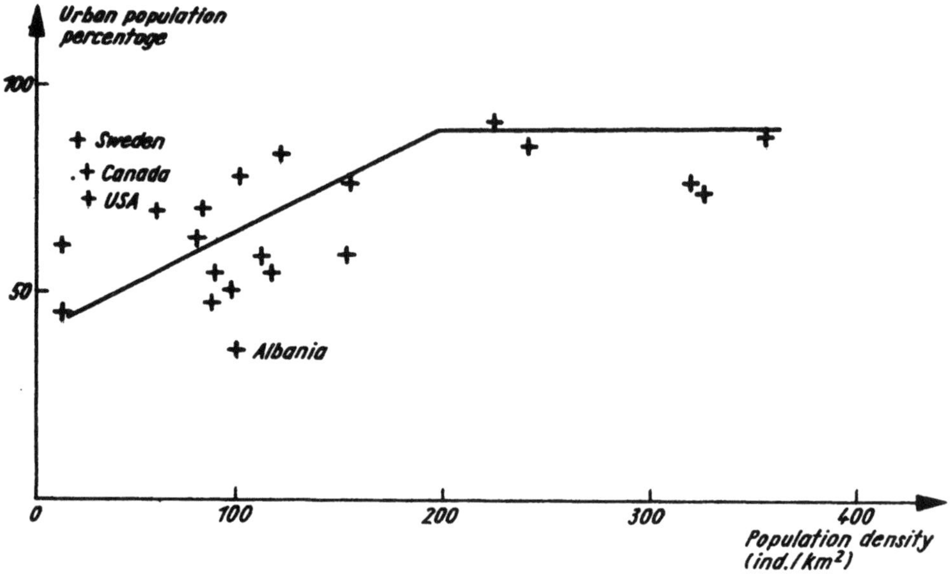

Fig. 25. Relationship between the urban population percentage and the average population density (corresponds to data of Table 3). Graph is described by the function $k(n) = \min\{k_0 + rn, k_a\}$, $k_0 = 0.4$, $k_a = 0.9$, $r = 0.003$ $(\text{ind./km}^2)^{-1}$

On the basis of the above we can infer that the population density after a nuclear exchange (n_v) as a function of population density before an exchange (n), can be represented as follows (cf. formula (5)):

$$n_v(n) = n(1-(1-\delta)k(n)) \qquad (7)$$

where k(n) is taken from (6). The graph of this function is shown in Figure 26. The breaking point of the curve corresponds to n = 200 ind./km², n_v = 56 ind./km², so that when $n > n$ ($n_v > n_v$):

$$n_v = 0.28\ n, \qquad (8)$$

and when $n < n$ ($n_v < n_v$):

$$n_v = 0.32n + 0.003\,n^2;$$

the inverse dependence therefore has the form:

$$n = -53.33 + \sqrt{(2844 + 333.3\,n_v)}. \qquad (9)$$

The second and main portion of the present subchapter is devoted to the conditions necessary for the outbreak of some epidemic diseases. This will be based largely on the results of the first part – it is assumed that for each concrete disease these conditions are determined solely by the density of the population surviving nuclear attacks and consequent disasters. We will assume that under the given conditions the level of medical care would fall sharply, so that an epidemic situation could be similar to that which existed in the Middle Ages during great epidemics of plague, cholera and other diseases. For model identification we will use data on the plague in Europe in the middle of the XIV century in accordance with the model proposed by Noble [110]. In a simplified version this model can be described as follows. Let n(t) be equal to the population

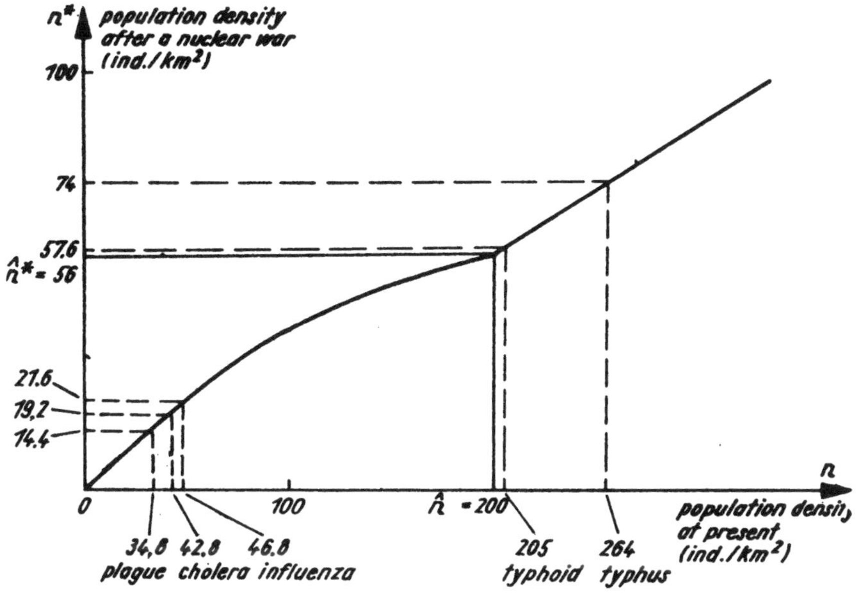

Fig. 26. Dependence of the population density after a war on the population density before a war. Critical values of population density are indicated for several epidemic diseases

density in a certain location at a given moment in time t; let S(t) be equal to the density of the infected subpopulation. The dynamic can be described by the following equation:

$$\frac{ds}{dt} = R(n-S)\,S - \mu S \tag{10}$$

The first term on the right-hand side describes the process of epidemic growth such that the increase in the number of sick individuals is proportional to the number of encounters between ill and healthy individuals. The second term describes a decline in the epidemic due to the recovery or death of sick individuals. The state of equilibrium S = 0 corresponds to the absence of epidemics. We assume that the epidemic will not spread if equilibrium is stable; the epidemic will spread if the equilibrium is unstable, i.e., for the inequality

$$n(t) > n_{cr} = \frac{\mu}{R}. \tag{11}$$

In the work cited above, a more complex scheme was adopted for modelling the plague, incorporating decrease in population density due to mortality among the infected population. However, neither this modification of the original model (10), nor any other, similar modification changes condition (11), under which the initial outbreak of the disease takes place. Finally, it should be noted that the means by which model (10) was constructed does not take into account the specifics of the disease; the model can therefore be used successfully in researching other diseases than the plague. The specific features of a given disease can be incorporated by changing the coefficients of proportionality R and μ in (10). As can be seen from (10), the rate μ is inversely proportional to the mean duration of a disease T_{sick} in one individual, or, more precisely, to the interval of time when the individual is contagious to other people. We now make one more hypothesis concerning the analogous dependence of the coefficient R on the mean incubation time T_{inc}. This means that the mechanisms of spreading for various diseases are such that the characteristic time of transmitting infection from one person to another is small in comparison with incubation time. This situation would be applicable, for example, under conditions of rather high population density. Rejecting this hypothesis would force us to use much more complex models, which would not allow for more or less reasonable identification, or else would be based on hypotheses that would be no more certain than the one proposed above. Thus we assume that for two different diseases $\mu^1 T^1_{sick} = \mu T_{sick}$, $k^1 T^1_{inc} = k T_{inc}$; using (11) we find:

$$n_{cr} = n^1_{cr} \frac{T^1_{sick}}{T^1_{inc}} \cdot \frac{T_{inc}}{T_{sick}} . \tag{12}$$

From Noble's data we can estimate n^1_{cr} = 14.4 ind./km² for the plague. Average incubation times and durations of disease are calculated from data taken from medical sources (eg., [14]). These data are contained in Table 4. For example, the average duration of plague is calculated on the basis of the average time until recovery (3–4 weeks) or until death (7 days), and from the percentage of fatal cases (40 %), giving us T^1_{sick} = 18 days.

The critical density values calculated by formula (12) are contained in Table 4. By substituting these values for n_v in formula (9), we can find the critical density values before a nuclear exchange, which are shown in the fourth column of Table 4.

Table 4. Characteristics of some contagious diseases

Disease	Mean incubation time (days)	Mean duration of a disease (days)	Critical population density ind./km²	Critical density before a war
Plague (bubonic)	3	18	14.4	34.8
Cholera	2	9	19.2	42.8
European typhus	12	14	74	264
Typhoid fever	14	21	57.6	205
Influenza	1	4	21.6	46.9

A comparison of these results with population density maps and with the data in Table 3 shows that diseases such as plague, cholera, and influenza would predominate in territories of Europe up to the Urals, the eastern part of North America, and the islands of Japan. For European typhus and typhoid fever the zone of spreading would be much smaller. Thus European typhus might be expected in Belgium, the Netherlands and Japan, while typhoid fever might also spread to such countries as Great Britain and the Federal Republic of Germany. It should be noted that this certainly does not mean that epidemics of these diseases would not appear in other countries.

Under the conditions following a catastrophe there always arise regions of extremely high population density. This means that various types of epidemic outbreaks would occur in such places. The present calculations can be considered to indicate the minimum level of possible infection of the population by contagious diseases. We have not included data on a number of other dangerous diseases with relatively long incubation periods. It is quite natural, however, to assume that the collapse of the existing structure of medical care would increase the probability of the outbreak and spreading of epidemic diseases such as tuberculosis, syphilis, hepatitis, etc.

10.5. Genetic consequences of a nuclear conflict on the human population

It should be remembered that *genetic* consequences (which are often harmless or almost harmless for the parents at the moment of determination by mutagenic factors) differ from somatic consequences in that their effects are prolonged under conditions of environmental stress. When passed on according to Mendel's laws of segregation, recombination, etc., genetic damage would lead, in all likelihood, to the appearance of various anomalies (from lethal to insignificant in phenotypic manifestation) among offspring, which would increase mortality and lower productivity in the population, and would lead to the appearance of socially inviable members in the society. Supporting such members of society would place a noticeable economic burden on the government, not to mention the moral burden on the parents of the afflicted.

10.5.1. The genetic effect of short-term acute irradiation

Let us first briefly discuss the genetic consequences of short-term acute irradiation for a population during a nuclear conflict. If the radiation dose were great, its effect would most likely be somatic, because of the sterilizing effect of radiation considered above. Although there can be no doubt that exposure to radiation can have pronounced mutagenic effects, the field of medical genetics does not have statistics available that would enable us to estimate precisely the genetic risk connected with irradiation [68]. Nevertheless, even for the highest doses under which productivity would be retained, the maximum risk of the appearance of hereditary anomalies is an order of magnitude lower than the risk of damage by one of the forms of hereditary anomalies already present in the population [68]. This means that the effect is substantially lower in comparison with somatic effects, even if we assume all anomalies to be lethal.

Of course, the total dose of body radiation received by the survivors of nuclear attacks varies greatly, but it is none the less insignificant in comparison with the usual scale of radiation genetics. The range of doses as a function of distance from the place of the blast and of effectiveness of protective barriers varies from zero to 600–700 R – the maximum survivable dose of total irradiation. Since the mean dose would be lower than this, the effects of irradiation on the first generation would be insignificant [56].

In order to elucidate the question of the role of irradiation, we can use data on the consequences of the nuclear bombs dropped on Hiroshima and Nagasaki. The genetic consequences of nuclear blasts would manifest themselves in a higher mortality rate in the offspring of irradiated persons, since they would be carriers of genes that had been altered by irradiation. Extended research on mortality in children whose parents survived the bombings of Hiroshima and Nagasaki was carried out on the following subjects [109]. Observations were made on:

1. 18,946 children, born alive, whose parents (one or both) were less than 2000 m from the hypocenter at the time of the blast, and together received a dose of 117 ber (one ber is the equivalent dose of radiation equal to 0.01 J/kg);
2 16,516 children whose parents were neither within 2000 m of the hypocenter nor received virtually any radiation dose;
3. 17,263 children whose parents were not in Hiroshima and Nagasaki at the time of the bombing. The average interval from birth to death or confirmation of survival was 17 years. An analysis of the data showed that the irradiation of the parents had no statistically significant influence on child mortality from various genetic defects.

The overall frequency of chromosomal anomalies among live births was 0.60 %. For children of parents who survived Hiroshima and Nagasaki, the corresponding frequency value is 0.62 %. Although this figure is only 0.28 % for the control group, the difference is not statistically significant [15].

Genetic damage as a result of irradiation would also have an effect on the productivity of the irradiated person. We now know that approximately 15 % of all pregnancies end in spontaneous abortions, while 2 % end in still births [7]. Much study has shown that approximately 47 % of spontaneous abortions are of a mutant (genetically anomalous) form, as are about 6.2 % of still births; for live births this percentage is only 0.6 % [10]. Thus genetic defects would lead to disturbances in the normal course of pregnancy, lowering the probability of live birth and, consequently, lowering productivity. Observations on women who survived the atomic bombing of Hiroshima and Nagasaki show that doses of 100 rad and above did not have a consistent effect on productivity [81].

A more detailed analysis of data on the dependence of birth defects and still births on doses of radiation in both parents has made it possible to confirm its presence and to calculate the "average" doubling dose (cit. [7]). The results of research on the genetic consequences of irradiation for data on residents of Hiroshima and Nagasaki is summarized in reports by the Scientific Committee on the Effect of Atomic Radiation (1977, 1978, 1980, 1982).

The absence of clear-cut genetic effects in the offspring of the irradiated population of Hiroshima and Nagasaki can be explained by the small size of these populations, the low yield of the nuclear explosions, the short duration of observation of the offspring (two generations), and the inadequacy of modern methods of exposing genetic defects [7].

10.5.2. The genetic effects of chronic irradiation

In an analysis of the consequences of nuclear radiation, we should take into account not only the effect of acute momentary irradiation, but also the influence of an increase in natural background radiation as a result of contamination of the environment, which leads to the cumulative accretion of the effects of weaker chronic irradiation over the course of a lifetime. According to our current understanding, any dose of radiation causes a proportional number of mutations (there is no threshold effect) having a negative influence [7]. In reports by the UN Committee on the Effect of Atomic Radiation [15, 31, 96], some estimates have been made of the risk of the appearance of genetic anomalies in man under the influence of radiation using indirect data (on the irradiation of embryonic cells of mice). Thus the general risk of induction of recessive mutations (which appear only when anomalous genes from the father and the mother meet at the same time in the organism of the offspring) is estimated to be $0.6 \cdot 10^{-4}$ per one gamete (sex cell) for one rad. Of the 60 mutations induced by irradiation of the sex cells in one million gametes, 1–3 mutations per million conceptions can be expected to appear in the first generation. This figure is based on data on drosophila. The general risk of induction of dominant mutations (which appear when one anomalous gene from either the father or the mother is present in the offspring) is estimated to be $20 \cdot 10^{-6}$ for one gamete at one rad (this estimate is based on data on mice). This figure should be seen as a rough approximation, and there is a wide range of opinions on this subject among geneticists.

The effect of irradiation is not limited to the induction of genetic mutations; it can also lead to chromosomal anomalies – more specifically to the induction of reciprocal translocations. The frequency of their appearance is estimated at $0.44 \cdot 10^{-4}$ per one gamete per rad. Given such frequencies, the consequences of irradiation of the fathers in a dose of one rad would be expressed as follows: 2–10 children with congenital defects, 11–55 cases of spontaneous abortions, and 22–109 cases of premature death of the fetus per million conceptions. These estimates are based on data on persons and monkeys.

The risk of an induction of loss of sex chromosomes is probably completely insignificant, considering that most of the corresponding conceptions are inviable.

The value of the double dose is also used in describing the effect of irradiation. If the level of mutations is doubled, the probability of bearing a diseased child is raised from 5 % to 10 % [10]. The use of the double dose is complicated by the fact that the dependence on dose is non-linear. This non-linearity is caused, for example, by the fact that large doses cause massive destruction and sterilization of sex cells, whereas the greater survivability of sex cells could allow even the weak effects of radiation to appear. However, there are statistics on the absence of a correlation between survivability and the appearance of genetic anomalies. The double dose for a person is estimated at 100 rad [15]. Under the assumption that the population was subjected to low-level radiation with a dose of one rad for a generation, it was calculated that in the first generation there would be 20 additional cases of offspring suffering from dominant diseases and diseases linked with the X-chromosome (sex chromosome), 40 cases of chromosomal diseases, and 5 cases of diseases of complex etiology per million live births (a total of 65 cases per million live births). This amounts to 6 % of the current frequency (10.7 %). At a level of equilibrium (which would be reached after various numbers of generations depending on the type of disease) this general increase would amount to 185 cases per million live births or 17 % of the current frequency [17].

According to the latest data, the general frequency of inherited pathologies is 10.5 %. Of these 1 % are dominant diseases or diseases linked with the X-chromosome, 0.1 % are recessive diseases (excluding those supported by the advantage of heterozygotes), 0.4 % are chromosomal diseases, and 9 % are congenital defects and other irregularly inherited diseases [29]. Let us consider the effects of various doses of radiation on these classes of inherited anomalies. First of all we should note that genetic consequences are largely dependent on the age of the irradiated persons. As an illustration we will introduce the following data [108] (see Table 5).

The effect of irradiation on various classes of diseases can be illustrated by the following data [29], contained in Table 6. According to Table 6, the increase in the frequency of inherited pathologies in the first generation would be on the order of 2 % of the current frequency of inherited anomalies; if each generation were irradiated with a dose of 100 rad, the number of inherited anomalies at equilibrium would rise to approximately 14 % of the current frequency. According to other data [7], these values would be 6 % and 17 %, respectively. Thus at equilibrium there would be an order of 12 % of inherited diseases in the population rather than 10.5 %. Such an increase of 1.5 % cannot be considered catastrophic.

Table 5. Number of genetic diseases as a function of age of irradiated persons

Age of irradiated persons (year)	Genetic diseases per 10^4 individuals per 1 Sv (= 100 rem)	
	Children and grandchildren of irradiated persons	Total number among all subsequent descendants of irradiated persons
0–30	100	200
18–30	100	200
entire population 0–73	40	80
18–65	26	52

Table 6. The effect calculated for 1 Gy/generation (1 Gy = 100 rad) of ionizing radiation with a low linear transfer of energy in a low dose or under low dose intensity for a population of 10^6 newborns using the method of double dose (double dose is assumed to be 1 Gy)

Classification of diseases		Increase in frequency of diseases	
	current frequency	effect of 1 Gy first generation*	generation equilibrium
Autosomal dominant diseases and diseases linked with the X-chromosome	10^4	1500	10^4
Recessive diseases (including those supported in a heterozygous state)	2500	weak	slow increase
Chromosomal diseases:			
structural	400	240	400
numerical	3000	probably very low	probably very low
Conganital anomalies, anomalies appearing in a late period, and constitutional and degenerative diseases	$90 \cdot 10^3$	450	4500
Total	105,900	2190	14,900

* It is assumed that the frequency in the first generation is equal to approximately 15 % of the frequency in equilibrium for autosomic, dominant and sex-linked diseases; approximately 3/5 for structural anomalies and 10 % for diseases of complex inheritance.

It is possible to explain the reason for the small increase in inheritea diseases using generally-accepted notions. The normal human genetic apparatus has become so finely-balanced in the process of evolution that any change of it will have an overwhelming probability for being harmful. Therefore the majority (50—70 %) of such changes leads to the impossibility of the development of the fetus, even in the earliest stages [10], i.e., to termination of the pregnancy. Such an outcome corresponds to the minimal "cost" of treatment of genetic anomalies from both a medical point of view and the point of view of the government, as compared with the birth of a child with a progressive illness, that might end in disability or death at a mature age. Only a very small number of changes would have a sufficiently low pathological impact in order for the pregnancy not to be terminated and to end in the birth of a viable but defective child. The tables were constructed assuming a weak dose with a low linear transfer of energy.

10.5.3. On the "cost" of the inherited burden on human society in connection with the consequences of a nuclear conflict

The social inferiority of members of a society with various inherited diseases, the economic burden that must be borne by the government due to the expenses involved in supporting the sick, and the heavy moral burden on the parents of affected children make any measures for lowering the level of inherited diseases of vital importance. In connection with this, methods of determining the level of mutagenic factors, including ionizing radiation, are being actively developed.

This aspect of the genetic consequences, along with the medical consequences [84] of a nuclear conflict is touched upon in various publications [7]. Efforts at developing indices of the burden of inherited anomalies have been summarized by the Committee on the Effect of Atomic Radiation [29]. As a measure of this burden we simply use the frequency of anomalies. However, an index of the number of inferior life years (including years spent under treatment, the period of time during which the subject is unable to work because of illness, etc.) is more accurate. By summarizing this type of data according to various classes of inherited diseases, distinguished by seriousness of the disease, time of appearance, etc., we can characterize loss of life years for the society due to genetic pathology. We will present corresponding data (cit. [29], Table 50) describing current estimates of the burden of genetic diseases (see Table 7).

Table 7. A comparison of estimations of the burden for various types of spontaneously arising genetic diseases

Category of disease	Frequency (per 10^6 births)	Average number of years of normal life	of inferior life	of lost life	Inferior years (per 10^6 births)	Lost years (per 10^6 births)
Autosomic dominant	$9 \cdot 10^3$	33	25 (33 %)[a)]	13	$225 \cdot 10^3$	$117 \cdot 10^3$
Linked with X-chromosome	10^3	1.5	40 (39 %)[a)]	28	$20 \cdot 10^3$[b)]	$14 \cdot 10^3$[b)]
Recessive	2500	1	35 (80 %)[a)]	32	87,500	$80 \cdot 10^3$[b)]
Chromosomal	$4 \cdot 10^3$	2	44 (55 %)[a)]	24	$176 \cdot 10^3$	$96 \cdot 10^3$
Irregularly inherited	$9 \cdot 10^3$	20	20[a)]	30	$18 \cdot 10^5$	$27 \cdot 10^5$
Total	106,500				2,308,500	3,007,000

a) percentage of inferiority
b) frequency of births used is lower by one half to account for both sexes

It should first be emphasized that estimates of the level (especially for equilibrium) of the inherited burden under the effect of ionizing radiation, as well as estimates of "loss of life years" were made under the assumption of the current living conditions (medical care, level of nourishment, social institutions, etc.).

Even under these conditions, maximum doses of acute and chronic irradiation cause an increase in genetic pathologies significantly lower than the value of the inherited burden in contemporary population. In comparison with the direct impact of other damaging factors on the population, this factor is undoubtedly of secondary significance.

However, in the absence of other damaging factors, genetic anomalies could be passed from generation to generation for a very long time if not counteracted by some sort of powerful forces. In this case these forces would be factors related to the deterioration of the external environment, which would become so extreme that it would border on the limits of what man can survive in general, depending on genetic heterogeneity.

It is well-known that the progress made by civilization, improvement of living conditions, and current medical advances have resulted in the fact that people with various diseases (including inherited diseases) who would previously have been doomed to death (or at least sterility) live to a reproductive age and bear offspring. This very fact increases the inherited burden on humanity. In addition, the burden on society of supporting those who cannot take care of themselves because of genetic anomalies increases, and the amount of resources that must be spent on them is rather substantial. It should be emphasized that this category of people can exist only on account of the well-known expenditures for their medical care, etc.

If the society were placed under extreme conditions, with a sharp deterioration in social institutions of medical care and other services, and if the very survival of people (who are completely normal) were called into question as a result of exceptionally strong environmental stress, the society would be unable to support individuals with genetic pathologies. In addition, climatic factors that are detrimental to the health, the effects of pollution, the destruction of agro-ecosystems, starvation and many other factors of environmental stress would place increased demands on normal human activity. Therefore individuals whose deviation from the norm under favourable conditions was practically insignificant would be subject to severe pressure from an environment that ruthlessly rejects all anomalies. In terms of genetics the given situation is the opposite of the situation being considered in connection with progress in medicine and health. Pathological genes would now be under much greater pressure of natural selection, which would lead to a sharp decrease in the level of the inherited burden. Consequently the indices presented here of the genetic burden induced by the effect of ionizing radiation (or other factors) should be considered to be maximum in scope; at the same time the demographic characteristics (mortality, fertility) of the population would become worse because of the pressure of natural selection on anomalous genetic variations, decreasing the probability of *survival* for people under unfavourable conditions after a nuclear war.

It is true that there exists still another genetic aspect of the situation following a nuclear war. This is the "insularization" of the population, i.e., the appearance of small groups within an originally large population; among these groups there would be virtually no interbreeding, while within them the degree of inbreeding would increase sharply. This could lead to an increase in the frequency of inherited diseases within each of these groups. At present it is difficult to analyze this effect quantitatively.

10.6. Conclusion

a) A global demographic catastrophe would erupt, as a result of somatic damage caused by the direct effects of nuclear weapons (thermic effects, shock wave, radiation), a sharp deterioration of environmental conditions ("nuclear winter" and "nuclear night", radioactive and "ordinary" pollution, increased level of UV-radiation), the destruction of natural and agricultural ecosystems, and the destruction of the social environment.

The negative impact of radioactive and "ordinary" pollution on human fertility, and an increase in the mortality rate due to extreme conditions would lead to a sharp decrease in the demographic potential of the human population and, as a result, to a negative population dynamic. Even those individual human populations that would survive a nuclear war could die out in the course of several generations.

b) After a nuclear war we could expect a significant decrease in the population density, which in itself would lower the probability of epidemic outbursts. However, the absence of medical care, a general deterioration in the state of health of the population and massive migrations would make this probability rather high. Diseases such as the plague and cholera would spread throughout all of Europe and North-East America. Outbreaks of other epidemic diseases would become more frequent. On the whole the epidemic situation would be quite similar to the situation in medieval Europe during the period of epidemics, and it would be also aggravated by the prolonged action of ionizing radiation.

c) The effect of acute irradiation as a result of nuclear attacks would be primarily somatic. The effect of chronic irradiation (as a result of radioactive contamination by long-lived isotopes) would lead to an average increase in the frequency of mutations and the number of genetic anomalies even in the first generations (following a nuclear war) by approximately 2 % of the current frequency (about 10 %). Due to extreme living conditions, anomalous variations would be subject to severe pressure from natural selection. On the one hand, such a situation would hinder an increase in the genetic burden on a population; on the other hand, demographic indices of fecundity and mortality would become worse, increasing the probability of the dying out of the population.

11. SIMULATIONS OF CONSEQUENCES OF A NUCLEAR WAR USING THE MODEL OF GLOBAL BIOSPHERE PROCESSES

11.1. Model of global biosphere processes created at the Computer Center of the USSR Academy of Sciences

The simulation of consequences of a nuclear conflict was based on a model of global biosphere processes (MGBP) [39] developed at the Computer Center of the USSR Academy of Sciences. The model is made of 3 interacting blocs: Land Regions, Ocean, Atmosphere. The land is divided into regions according to natural and political frontiers. The bloc Land Regions represents biogeochemical cycles of carbon and nitrogen for natural and agricultural vegetation, CO_2 exchange with the atmosphere, demographic and ecological processes. The land surface in the model is split into sections, 5 along geographical longitude by 4 along geographical latitude. Ecologically every section is specified by the amount of carbon and nitrogen in living and dead phytomass, in litter and in humus.

The bloc Ocean accounts for the CO_2 exchange between the ocean and the atmosphere, for the pollution of the ocean, and for the dynamics of biota. The bloc Atmosphere simulates the climatic parameters affected by man.

MGBP simulates long-term variations in ecological climatic and demographic processes in separate regions of the Earth for different initial biosphere parameters and scenarios of human activity (specific time is a decade). The model is realized as a computer program and provides the naturalist with a possibility to specify various scenarios of impact upon the biosphere in order to analyze their outcome. For instance, this model was used to calculate variations in global temperature and CO_2 content in the atmosphere after the "nuclear winter" (presented in Chapter 7), as well as in simulations of the vegetation recovery on land.

The model can simulate alterations in biological processes caused by multiple factors of different value, including very strong effects. However, it cannot represent practically an instantaneous death of plants and animals as a result of such a strong impact as caused by the radioactive effect of the nuclear war or the "nuclear winter".

Therefore, specially designed mathematical models (e.g. for estimating variations in the diversity of species) were used in the estimation of vegetation and animal mortality caused by the effect of the nuclear war factors. Extinction of vegetation due to irradiation and "nuclear winter" has been extrapolated from available experimental data and expert estimations. Fairly complicated mathematical procedures have been used in processing experimental data.

11.2. Scenario

The global model can calculate the dynamics of global and regional variables for a given scenario. In this variant of the model 7 regions will be considered:
1. Europe; 2. North America; 3. The Asian part of the USSR; 4. South America; 5. Asia (excluding the USSR); 6. Africa; 7. Australia.

Let us consider the following scenario:

a) It is assumed that territories of regions 1, 2 and 3 would be subjected to nuclear attacks.

b) The population in these regions would instantly fall by 80 %.

c) Mortality of humans and animals in these regions would increase 4 times (for humans this would be equal to the mortality rate of the Middle Ages). The human birth-rate and animal fertility in regions 1–3 would be decreased 3 times.

d) In regions 1–3 there would be an absence of capital and hence no investments.

e) The exchange of capital, food, fertilizers, and other resources among different regions would be completely cut off.

f) In regions 1–3, 10 % of all mixed forests would be destroyed immediately, and coniferous forests would be completely destroyed ("radiation shock", radioactive and "ordinary" pollution fires).

g) 70 % of agro-ecosystems would be wiped out in regions 1–3, while the productivity for those remaining would be 30 times less than the prewar level.

All these assumptions are based on estimates made in the previous Chapters.

The parameter we will vary will be the increase in the total level of environmental pollution connected mainly with radioactivity. In our scenario we will use the following assumption for global pollution of the biosphere:

h) Pollution of the biosphere would instantly increase k times in region 1–3 and k/5 times for the remaining regions. After this the coefficient of pollution generation in region 1–3 would become equal to zero, since the major source of anthropogenic pollution – industry – would be completely destroyed.

And, finally, the following assumption:

i) Fishing as a source of protein would be completely absent in region 1–3.

Calculations for various values of k were carried out according to this scenario using our global model.

11.3. Results of calculations

The value k = 3, i.e. when global pollution of the biosphere instantly increases 3 times, is critical. What does this mean? In our model there is a gradual flow of anthropogenically caused pollution into the environment (annual norm of pollution). Here this flow increases 3 times and it is considered to be the flow of radioactive contamination.

The dynamics of population density in each of the regions is represented in Figure 27. A decline in population density would begin after the war in all regions, independently of whether they were involved in the nuclear conflict or not. After 40 years, densities would rise somewhat, evidently as a result of a weakening in the impact of pollutants (mainly radioactive contaminants), but would then fall again, since other factors limiting the growth of human population would take on primary importance (mainly lack of food). Further calculations show that a transitional regime would begin after 40 years, and populations, though still oscillating, would very slowly begin to be restored.

The agricultural systems graph shows the percentage of decrease in agricultural productivity in regions not directly involved in the nuclear conflict. We can see that productivity falls, but after 30 years the decline would cease and a transitional process would begin. Further calculations show that after the transitional process (about 30 years), a slow increase in productivity would begin.

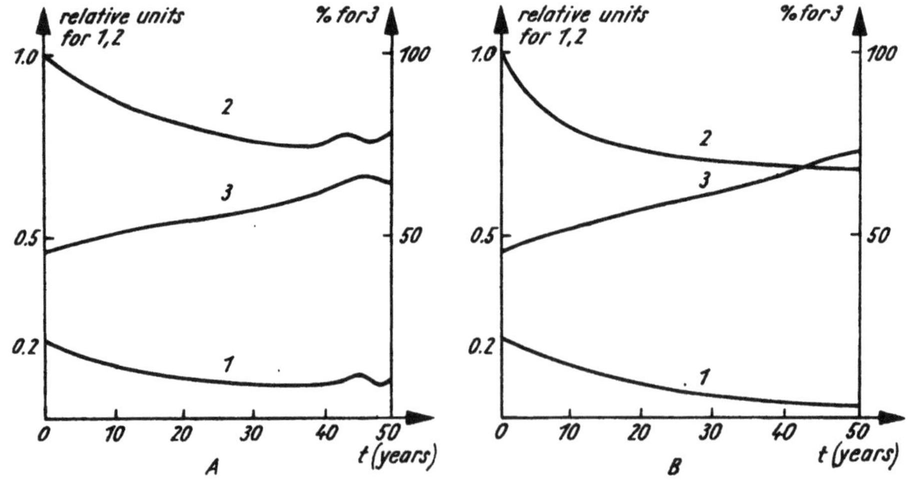

Fig. 27. Dynamics of population density and agricultural production after a nuclear war: 1 – population density for the regions 1 to 3; 2 – the same for the regions 4 to 7; 3 – decrease in agricultural production of the regions 4 to 7; A – contamination is increased K=3 times; B – K=5 times

(The graph indicates the percentage of decline in productivity; therefore an increase on the graph corresponds to a decline.)

Figure 27b shows the dynamics of the same process but for k=5. We can see that this is a degradation dynamics: the population would decrease, agricultural productivity would decline, etc.

We have not indicated the dynamics of agricultural productivity in the regions involved in the nuclear conflict. Since primitive agriculture would prevail in those territories, agriculture would be dependent on weather conditions and variations in climate.

It should be noted that pollution of k=3 is given the AMBIO scenario [78], and k=3 is the bifurcation value for the survival of human populations.

Since factors of the nuclear conflict have a cumulative effect, biosphere may lose stability at even lower values of these factors. The loss of stability as a process is yet more dangerous since it is practically impossible to forsee the further system evolution after it passes the bifurcation value.

11.4. Conclusion

In this chapter we have presented only one variant of simulation of the global dynamics after the nuclear conflict, based on a rather "tough" scenario. We may debate about this scenario, we may substitute it by a "softer" one, but the qualitative pattern of degradation is most likely to be preserved. In future we plan to perform a series of simulations according to different scenarios. The simulation discussed above is only the first stage of work which already seems rather impressive. It would be most helpful to discuss the possible scenarios for the new series, so that they would represent different viewpoints and approaches towards this important problem.

12. CONCLUSION

We have estimated each of the factors of the nuclear conflict separately. Each of these factors results in serious infractions of ecological processes on vast territories and eventually may bring ecosystems to complete destruction. What would be the synergetic impact of these factors? We have also considered the recovery of ecosystems affected by some of the factors of a nuclear war. How would the ecosystems recover after the joint action of these factors and against the background of the joint action of factors of the aftermath of the nuclear war?

The joint action of any two factors, all the more all of them, would be far more disastrous than the sole action of one factor. For instance, both the factors "radioactive contamination" and "nuclear winter" would disturb cell processes in plants and consequently their joint effect would be stronger. Quantitative estimations are extremely difficult to make.

In case of "UV-radiation" acting together with "radioactive contamination", it suppresses processes of cell repair, considerably increasing the radiosensibility of organisms and, hence, the level of somatic mutations. The suppression of the immune system in animals would be also much enhanced.

The transfer of mercury into fish organisms depends upon pH of water. The lower the pH, the higher is the intensity of mercury accumulation. Thus, in polluted water with pH=7.5, after a certain period of time, fish accumulate 0.4 mg/kg of mercury, for pH=5 the value is 2.7 mg/kg [50]. So the joint action of "acid rain" and "environmental pollution by heavy metals" may result in considerable contaminations of fish.

While separate factors may cause only damage to plants, their joint action may kill them completely. For example, in grasslands radioactive contamination can make seeds sterile, fires burn the vegetation, acid soils inhibit soil microorganisms. As a result, ecosystems may die completely. Their natural recovery (without introduction of species) would be impossible.

While acid rainfall in tropic forests of the Southern hemisphere slightly changes the intensity of soil processes, and the factors of the "nuclear winter" decrease productivity of plants, the joint action of these two factors in some cases may destabilize this fragile ecosystem and destroy it.

Similarly, due to the action of all factors of the nuclear conflict upon man, the probability of extinction of the human population becomes many times increased.

The enhancing effect of the joint action of factors would be most prominent in the case of the recovery of ecosystem. The recovery period would be many times increased, recovery may even become impossible. For instance, (see Chapter 7), according to estimations of the recovery process in coniferous forests of the temperate zone, if there were no soil erosion, in 50 years of recovery biomass of forest vegetation would be 41 % of the initial value. If the soil were destroyed by 50 %, after 50 years biomass would recover to 18 %, and if soil degradation were 75 %, the recovery of biomass after 50 years would be only 8 %.

The recovering forest, affected by the consequences of the nuclear conflict (radioactive soil contamination, extinction of birds, etc.), is likely to become invaded by pest insects and die again. In other words, ecosystems affected by consequences of a nuclear war would become low-stable with respect to extreme situations. As a result, ecosystem recovery would be significantly retarded.

The situation for human population would be similar. As already noted, human population may decline to extinction in the course of several generations after the end of the nuclear conflict. Even if recovery takes place, it would be very slow under the joint action of negative factors.

Restoration of industry and agriculture would be also very slow. The industrial restoration would be retarded by the need of most of the labour resources in food production under conditions of primitive agriculture. Restoration of agriculture would be impeded because of the ruined industry.

The main conclusion from our study is that even the most "optimistic" scenarios of the consequences of the nuclear conflict (if it is fair to speak of optimism in this case) would inevitably result in a global ecological and demographic crisis.

LITERATURE

[1] Aleksakhin, R.M.: Istoriia lesnoi radioekologii, ee dostizheniia i nekotorye nereshennye zadachi. In: Problemy lesnoi radioekologii, Gidrometeoizdat, Moscow 1979, pp. 6–26.
[2] Agroklimaticheskji atlas Mira. Moscow 1972.
[3] Aleksandrov, A.P. (ed.): ladernaja energetika, chelovek i okruzhaiushchaia sreda. Energoizdat, Moscow 1981.
[4] Aref'eva, Z.N.: Vliianie ognia na nekotorye biokhimicheskie protsessy v lesnykh pochvakh. Trudy Instituta biologii UFAN SSSR, vypusk 36, 1963, pp. 39–57.
[5] Armand, A.D., Vediushkin, M.A., Krapivin, V.F., Svirezhev, Y.M., Tarko, A.M.: Matematicheskaia model' vozdeistviia zagriaznenii na lecnoi biogeotsenoz. In: Tezisy dokladov vsesoiuznoi shkoly "Vliianie promyshlennykh predpriiatii na okruzhaiushuiu sredu", Pushchino, 1984, p. 13.
[6] Atlas USSR, Moscow 1983, p. 100.
[7] Baev, A.A., Bochkov, N.P.: ladernaia voina postavit pod somnenie sushchestvovanie cheloveka kak biologicheskogo vida. Priroda, 1983, No. 10, 3–10.
[8] Bazilevich, N.I.: Biogeochimiia glavneishikh tipov rastitel'nosti Zemli. In: Tezisy dokladov delegatov u s''ezda Vsesoiuznogo botanicheskogo obshchestva, Kiev 1973, pp. 239–244.
[9] Baranov, V.F.: Dozimetriia elektronnogo izlucheniia. Atomizdat, Moscow 1974.
[10] Bochkov, N.P.: Mutatsionnyi protsess u cheloveka i prognozirovanie ego effektov. Priroda, 1981, No. 2, 32–39.
[11] Brachina, A.N., et al.: Metodika i resul'taty izucheniia samoochishcheniia vody vodoemov, zagriaznennykh smes'iu radioprotonov. In: Metody radioekologicheskikh issledovanii, Atomizdat, Moscow 1971, pp. 113–123.
[12] Vakurov, A.D.: Lesnye pozhary na Severe. Nauka, Moscow 1975.
[13] Verkhovskaja, I.N. (ed.): Metody radioekologicheskikh issledovanii. Atomizdat, Moscow 1971.
[13a] Veinberg, B.P.: Ljod. Moscow, Leningrad, 1940.
[14] Voikulesku, M. (ed.): Infektsionnye bolezni. Meridian, Bukharest 1964.
[15] Geneticheskie posledstviia radiatsii. Nauchnyi komitet OON po deustviiu atomnoi radiatsii. Dokument A/AC. 345, 25 January, 1977, prilozhenie N.
[16] Girs, G.I.: Fiziologiia oslablennogo dereva. Nauka, Novosibirsk 1982.
[17] Guliaev, B.I.: Ocherk napravlennoi ekofiziologii. Itogi nauki v tekhniki VINITI Botanika. 1984, v. 5, 3–138.
[18] Guniazhenko, I.V.: Vliianie nizovykh pozharov na nekotorye agrokhimicheskie svoistva pochv sosnovykh zherdniakov. In: Lesovedenie i lesnoe khozieistvo. Vypusk 4, Minsk 1971, pp. 64–70.
[19] Gushchin, G.P., Vinogradova, N.N.: Summarnyi ozon v atmosfere, Gidrometeoizdat, Leningrad 1983.
[20] Grechushkina, M.P.: Tablitsa sostava produktov mgnovennogo deleniia U^{235}, U^{236}, Ru^{239}. Moscow 1964.
[21] Gusev, N.G., Dmitriev, P.P.: Kvantovnye izluchenie radioaktivnykh nuklidov. Atomizdat, Moscow 1977.
[22] Deistvie iadernogo oruzhiia. Ed. E. Gleston (transl. from English). Voenizdat, Moscow 1966.
[23] Dubrov, A.P.: Deistvie ul'trafioletovoi radiatsii na rasteniia. Izd. AN SSSR, Moscow 1963.
[24] Dubrov, A.P.: Geneticheskie i fiziologicheskie effekty deistviia ul'trafioletovoi radiatsii na vysshie rasteniia. Nauka, Moscow 1968.
[25] Zaozerskii, I.N.: Neorganicheskaia khimiia. Vysshaia shkola, Moscow 1963.
[26] Ivanov, A.I.: Porazhaiushchee deistvie iadernogo vzryva. Voenizdat, Moscow 1960.
[27] Ivanov, V.N.: Kurs dozimetrii. Atomizdat, Moscow 1978.
[28] Izrael', Iu.A., Stukin, E.D.: Gamma-izluchenie radioaktivnykh vypadenii. Atomizdat, Moscow 1967.
[29] Ioniziruiushchee izluchenie: istochniki i biologicheskie effekty. Nauchnyi komitet OON po deustviiu atomnoi radiatsii. Doklad za 1982 na General'noi Assamblee, v. II, New York, UN, 1982.
[30] Isaev, A.S., Girs, G.I.: Vzaimodeustvie dereva i nasekomykh kislofagov, Novosibirsk 1975.

[31] Istochniki i destvie ioniziruiushchei radiatsii. Nauchnyi komitet OON po destviiu atomnoi radiatsii. Doklad za 1977 na General'noi Assamblee, v. III, New York, UN, 1978.
[32] Karaban, R.T., Mishenkov, N.N., Prister, B.S., Aleksakhin, R.M., Tikhomirov, F.A., Fedorov, V.A., Romanov, G.N.: Deistviia ostrogo γ-oblucheniia na lesnoi biogeotsenoz. In: Problemy lesnoi radioekologii. Gidrometeizdat, Vypusk 38, Moscow 1979, pp. 27–52.
[33] Kasatkina, V.V.: Zarubezhnye atomnye elektrostantsii. Tsniiatominform, Moscow 1979.
[34] Kasperska-Palach, A.: Mekhanizm zakalivaniia travianistykh rastenii. In: Samygin, G.A. (ed.): Kholodostoikost' rastenii. Kolos, Moscow 1983, p. 112.
[35] Kaurichev, I.S. (ed.): Pochvovedenie. Kolos, Moscow 1975.
[36] Kashcheev, A.L.: Zabolachivanie vyrubok i mery bor'by s nim. Izd. AN SSSR, Moscow 1955.
[37] Kolobashkin, V.M., Rubtsov, P.M., Aleksakhin, V.G., Ruzhanskii, P.A.: Beta-izluchenie produktov deleniia. Atomizdat, Moscow 1977.
[38] Kolobashkin, V.M., Rubtsov, P.M., Ruzhanskii, P.A., Sidorenko, V.D.I.: Radiatsionnye kharakteristiki obluchennogo iadernogo topliva. Energoatomiazdat, Moscow 1983.
[39] Krapivin, V.F., Svirezhev, Y.M., Tarko, A.M.: Matematicheskoe modelirovanie global'nykh biosfernykh protsessov. Nauka, Moscow 1982.
[40] Kudriashov, Y.B., Berenfel'd, B.S.: Osnovy radiatsionnoi biofiziki. Izdat. Mosk. Universiteta, Moscow 1982.
[41] Kuleshova, L.V., Potapova, N.A., Pylenkova, L.V.: Opyt kompleksnogo izlucheniia dinamiki soobshchestv na gariakh v usloviiakh okhraniaemoi territorii. In: Prirodnye zapovedniki i osnovnye printsigy ikh raboty. Minsk 1977, pp. 144–147.
[42] Kulikov, N.V.: Radioekologiia presnovodnykh rastenii i zhivotnykh. In: Sovremennye problemy radiobiologii. Vol. 2 Radioekologiia, Atomizdat, Moscow 1976, pp. 369–374.
[43] Kurbatskii, N.P.: Tekhnika i taktika tusheniia lesnykh pozharov. Goslesbumizdat, Moscow 1962.
[44] Kurbatskii, N.P.: Problema lesnykh pozharov. In: Vozniknovenie lesnykh pozharov. Nauka, Moscow 1964.
[45] Larkher, V.: Ekologiia rastenii. Mir, Moscow 1978.
[46] Lishtvan, I.I., Korol', N.T.: Osnovnye svoistva torfa i metodika ikh opredeleniia. Minsk 1975.
[47] Margiulenene, D.P., Polikarpov, G.G.: Puti postupleniia radionuklidov v presnovodnye molliuski i lichinki nasekomykh. Radiobiologiia (1969) No. 2.
[48] Mashkovich, V.I.: Zashchita ot ioniziruiushchikh izluchenii. Energoatomizdat, Moscow 1982.
[49] Melekhov, I.S.: Lesnaia pirologiia. Lesnaia promyshlennost', Moscow 1978.
[50] Mel'nikov, N.N., Volkov, A.I., Korotkova, O.A.: Pestitsidy i okruzhaiushchaia sreda. Khimiia, Moscow 1977.
[51] Molchanova, I.V., Karavaeva, E.I., Kulikov, N.V., Chebotina, M. Ia.: Raspredelenie ^{90}Sr and ^{137}Cs po komponentam bolotno-rechnoi sistemy. Ekologiia (1982) No. 2, 45–49.
[52] MEMO. Ekonomicheskoe polozhenie kapitalisticheskikh i razvivaiushchikhsiz stran. Obzor za 1983 g. i nachalo 1984 g. "Pravda", Moscow 1984.
[53] Nazim, A., Dzheims, A.: Zhizn' v usloviiakh intensivnogo oblucheriia. In: Kashner, D. (ed.): Zhizn' mikrobov v ekstremal'nykh usloviiakh. Mir, Moscow 1981, pp. 470–504.
[54] Nasyrov, Y.S., Asrorov, K.A., Pinkhasoy, Y.I.: Fotosintez khlopchatnika. In: Fiziologiia khlopchatnika. Kolos, Moscow 1977, pp. 88–103.
[55] Nesterov, V.G.: Pozharnaia okhrane lesa. Goslestekhizdat, Moscow 1945.
[56] Neil, D., Shell, U.: Nasledstvennost' cheloveka (transl. from English). IL, Moscow 1958.
[56a] Normi radiatsionnoi besopasnosti. Atomizdat, Moscow 1978.
[57] Oksegendler, G.I.: Iady i protivoiadiia. Nauka, Leningrad 1982.
[58] Parchevskaia, D.S.: Otsenka pogloshchennykh doz β-izluchenii v sharovichykh gidrobiontakh melkogo razmera. Radiobiologiia (1969) v. 2, 381–385.
[59] Popova, E.P.: O prodolzhitel'nosti pirogennogo vozdeistviia na svoistva lesnykh pochv. In: Gorenie i pozhary v lesu. Krasnoiarsk 1978, pp. 185–186.
[60] Prokhorov, V.M., Safronova, N.G.: Kinetika samoochishcheniia vodoema, soderzhashchego strontsii-90 v rezul'tate pogloshcheniia radionuklido dannymi otlozheniiami. Ekologiia: (1973) No. 2, 11–18.
[61] Ramad, F.: Osnovy prikladnoi ekologii. Gidrometeoizdat, Leningrad 1981.
[62] Rovinskii, F.L., Makhon'ko, K.P.: K voprosy o migratsii radioaktivnoi primesi v gruntakh neprotochnykh vodoemov. Trudy I.P.G. 8 (1967) pp. 64–69.

[63] Safronova, N.G., Pitniaken, G.B., Pogodin, R.I.: O mekhanizmakh migratsii ^{90}Sr, v donnykh otlozheniiakh vodoemov. Trudy instituta ekologii rastenii i zhivotnykh DNTs AN SSSR **110** (1978), 95–99.
[64] Sivkov, S.I.: Metody rascheta kharakteristik solnechnoi radiatsii. Gidrometeoizdat, Leningrad 1968.
[65] Svirezhev, Y.M., Krapivin, V.F., Tarko, A.M.: Ekologicheskie posledstviia iadernogo konflikta. SCOPE – ENUWAR Meeting. Leningrad 1984.
[66] Sofronov, M.A., Vakurov, A.D.: Ogon' v lesu. Nauka, Novosibirsk 1981.
[66a] Stebljuk, I.I.: Zaschita lesnykh nasazhdenij ot massovogo porazheniya. Izd. Acad. selsk. hoz. USSR, 1983.
[67] Stenchikov, G.L.: Posledstviia iadernoi voiny: Vybrosy i rasprostranenie opticheski aktivnykh primesei v atmosfere. Vychislitel'nyi tsentr AN SSSR, 1985.
[68] Stevenson, A., Davison, B.: Mediko-geneticheskoe konsul'tirovanie (transl. from English). Mir, Moscow 1972.
[69] Sukachev, V.N.: Bolota, ikh oprazovanie, rasvitie, i svoistva. In: Isbrannye trudy, vol. 2, Moscow/Leningrad 1973.
[70] Timofeeva-Resovskaia, E.A.: Raspredelenie radionuklidov po osnovnym komponentam presnovodnykh vodoemov. Trudy instituta biologii UFAN SSSR **39** (1963), 23–48.
[71] Tikhomirov, F.A.: Deistvie ioniziruiushchikh izluchenii na ekologicheskie sistemy. Atomizdat, Moscow 1972.
[72] Tumanov, I.I.: Fiziologiia zakalivaniia i morozostoikosti rastenii. Nauka, Moscow 1979.
[73] Tiuremnov, S.N.: Torfianye mestorozhdeniia. Moscow/Leningrad 1949.
[74] Shipunov, F.Ia., Stepanov, A.M., Frolov, V.A.: Zagriaznenie biosfery v Severnom polusharii na fonovom urovne. In: Antropogennye narusheniia i prirodnye izmeneniia nazemnykh ekosistem. AN SSSR, Moscow 1981, pp. 7–28.
[75] Shostakovich, V.B.: Lesnye pozhary v Sibiri v 1915. In: Ocherki po zemledeliiu Vostochnoi Sibiri, vol. 47, Irkutsk 1924.
[76] Agriculture: Toward 2000. FAO, Rome 1981.
[77] Alexandrov, V.V., Stenchikov, G.L.: On the modelling of climatic consequences of the nuclear war. Computer Center USSR Acad. Sci., Moscow 1983.
[78] Ambio 11 (1982), 2–3.
[79] Barnaby, F., Rotblat, J.: The effects of nuclear weapons. Ambio **11** (1982) 2–3, 149–152.
[80] Barnaby, F., Kristoferson, L., Peterson, J., Prawitz, J., Rodhe, H., Rotblat, J.: The Reference Scenario: How a Nuclear War Might Be Fought. Ambio **11** (1982) 2–3, 94–99.
[81] Blot, W., Savada, H.: Fertility among female survivors of the atomic bombs of Hiroshima and Nagasaki. Amer. J. Hum. Genet. **24** (1972) 6, 613–622.
[82] Bolin, B., Degens, E.T., Kempe, S., Kethner, P. (eds.): The global carbon cycle. SCOPE 13. Wiley & Sons, New York 1979.
[83] Cline, M.G., Salisbury, F.B.: Effect of UV-radiation on leaves of higher plants. Rad. Bot. **6** (1966) 2, 151.
[84] Coggle, J., Lindop, P.: The medical consequences of radiation following a global nuclear war. Ambio **11** (1982) 2–3, 106–113.
[85] Crutzen, P.J., Heidt, L.E., Krasnec, J.P., Pollock, W.H., Seiler, W.: Nature **282** (1979), 253.
[86] Crutzen, P., Birks, J.: The atmosphere after a nuclear war: twilight at noon. Ambio **11** (1982) 2–3, 114–125.
[87] Deleape, J.P., Julien, J.M., Sauget-Naudin, N., Souchon, C.: Eco-energetics analysis of an agricultural system: the French case in 1970. Agro-Ecosystems (1979) 5, 345–365.
[88] Dochiger, L.S., Bender, F.W., Fox, F.L., Heck, W.W.: Chlorotic dwarf on eastern white pine caused by an ozone and SO_2 interaction. Nature **225** (1970), 476.
[89] Effects of ionizing radiation on aquatic organisms and ecosystems. International Atomic Energy Agency. Technical report series 172, IAEA, Vienna 1976.
[90] Ehrlich, P.R., Harte, J., Harwell, M.A., et al.: Long-term biological consequences of nuclear war. Science **222** (1983), 1293–1300.
[91] Evans, L.F., Weeks, I.A., Eccleston, A.J., Packham, D.R.: Environmental Science and Technology **11** (1977) 896.
[92] FAO fertilizer yearbook, 1982. FAO, Rome 1983.

[93] FAO production yearbook, 1982. FAO, Rome 1983.
[94] FAO trade yearbook, 1982. FAO, Rome 1983.
[95] FEMA, What the planner needs to know about fire ignition and spread. Federal Emergency Manag. Agency, Attack Environment Manual, CFG, Ch. 3, 2–1A3, May 1982.
[96] Genetic effects of radiation. UN Scientific Committee on the Effect of Atomic Radiation. A/AC.82/R.380, 10 June, 1980.
[97] Hatano, S., Kabata, K.: Transition of lipid metabolism in relation to frost hardiness in Chlorella Ellipsoida. In: Zi, P.H., Sakai, A. (ed.): Plant cold hardiness and freezing stress mechanisms and crop implications. V.2. Academic Press, New York 1982, pp. 145–156.
[98] Hill, J.E.: Problems of fire in nuclear war. Physical Department RAND Corp. P 2414, 1961, p. 32.
[99] Hollingsworth, T.H.: A demographic study of the British ducal families. In: Glass, D.V., Eversley, D.E.C. (ed.): Population in history. Essays in historical demography. Arnold, London 1965, pp. 354–378.
[100] Jørgensen, S.E. (ed.): Handbook of Environmental Data and Ecological Parameters. 1978. Vaerdøse. Denmark ISFM, p. 1162.
[101] Kangas, P.: Simulating succession as a reclamation alternative. In: Anal. Ecol. Syst.: State A Ecol. Modell., Amsterdam 1983, pp. 451–455.
[102] Koller, Z.R.: Ultraviolet Radiation (2nd ed.). Wiley, New York 1965.
[103] Kuusi, P.: This world of man. Werner Söderstrom, Helsinki 1982.
[104] Laulan, Y.: Economic consequences: back to the dark ages. Ambio 11 (1982) 2–3, 149–152.
[105] Larson, B.A., Small, R.D.: Analysis of the large urban fire environment. Part II. Parametric analysis and model city simulations. PSR 210, Pacific Sierra Research Corporation, Los Angeles, CA, USA, 1982.
[106] Larsson, A.: Some biochemical effects of cadmium on fish. Sublethal Eff. Toxic Chem. Aquatic Animals, Amsterdam 1975.
[107] Marr, J.W.: Ecology of the forest-tundra ecotone on the east coast of Hudson Bay. Ecolog. Monogr., v. 18, N 1, 1948, pp. 117–144.
[108] Myers, D.: Low level radiation: a review of current estimates of hazards to human populations. Chalk River Nuclear Laboratories, Chalk River, Ontario, 1982.
[109] Neel, J., Kato, H., Schull, W.: Mortality in the children of atomic bomb survivors and controls. Genetics 76 (1974) 2, 311–326.
[110] Noble, J.V.: Geographic and temporal development of plagues. Nature 250 (1974), 726–729.
[111] Principles of Ecotoxicology. SCOPE 12. J. Wiley & Sons, New York 1979.
[112] Richard, W.H., Sauer, R.H.: North-West Sci. 56 (1982) 1, 41–47.
[113] Seiler, W., Crutzen, P.J.: Climate change, v. 2, 1980.
[114] Schroeder, M.J., Chandler, C.C.: Monthly fire behavior patterns. US Forest Service Research Note PSW-112, 15, 1966.
[115] Smil, V., Nachman, P., Long, T.V.: Energy analysis and Agriculture. Westview Press, Boulder, Colorado, 1983.
[116] Statistical Abstract of the USA, 1982–1983. US Dept. of Commerce, Washington 1984.
[117] Svirezhev, Y.M., Krapivin, V.F., Tarko, A.M.: Ecological consequences of nuclear conflict. Meeting SCOPE-ENUWAR, Leningrad, May 1984.
[118] Taylor, O.C., Dugger, Jr., W.M., Cardiff, E.A., Darley, E.F.: Interaction on light and atmospheric photochemical (smog) within plants. Nature 192 (1981), 814.
[119] Turko, R.P., Toon, O.B., Ackerman, T., Pollack, J.B., Sagan, C.: Nuclear winter: global consequences of multiple nuclear explosion. Science 222 (1983), 1283–1292.
[120] Westing, A.H.: Weapons of Mass Destruction and the Environment. Taylor and Francis, 1977.